FUNDAMENTAL CONCEPTS OF INFORMATION MODELING

FUNDAMENTAL CONCEPTS OF INFORMATION MODELING

BY MATT FLAVIN

Yourdon Press
1133 Avenue of the Americas
New York, New York 10036

This book was set in Times Roman by YOURDON Press, 1133 Avenue of the Americas, New York, N.Y., using a PDP-11/45 running under the UNIX[†] operating system.

[†]UNIX is a registered trademark of Bell Laboratories.

Contents

FUNDAMENTAL CONCEPTS OF INFORMATION MODELING

Preface

Information modeling is a modern form of system analysis that identifies the objects, relationships, and operations composing some real-world system. It is used for database design and business system analysis and planning.

As an analytical procedure, it is composed of two major parts: an analytical "front-end," and a representational "back-end." The analytical front-end is a coherent set of procedures for finding, identifying, and defining objects; relationships; operations that modify the objects and relationships; and data elements that describe objects and relationships. The representational back-end is a set of procedures for mapping the semantic components of the model onto data structures that represent and describe each component. Information modeling is a marriage of the art of system analysis with the science of data representation.

This monograph is intended to make the system analysis component less artistic and a bit more scientific, and to lay out a proper conceptual foundation for the construction of an entity-relationship (E-R) model of some real-world system. Three fundamental aspects are studied in detail:

- theory of objects (entities)
- theory of relationships
- unity of the entity-relationship model

The purpose of this book is to introduce the reader to an explanation of what information modeling is, and to introduce innovations in model semantics and analytical procedures that form the basis of information modeling. *Fundamental Concepts of Information Modeling* is not intended to be a full treatment of the subject, but is intended to be a concise technical primer of the basic innovations in the theory and techniques of information modeling.

Information modeling differs in a very significant way from the prominent methods of semantic information analysis that grew up during the 1970s. Those methods were based on the formal analytical ideas of Ted Codd of IBM. I will group these methods under the name

1

"mathematical dependency analysis." Taken by itself, mathematical dependency analysis has great value; but it has been misapplied as a database design procedure. It is a fundamental premise of information modeling that before any kind of mathematical dependency analysis can have meaning, *it is necessary that the rules, conventions, and laws that govern the structure and operation of the real-world system be understood and articulated.*

The central method of attack on the problem of uncovering and articulating basic business policy is entity-relationship analysis. Each entity is a focus of business policy, and relationships between entities are clearly policy-dependent. Once we know what the underlying policy (rules, laws, conventions) really is, then we can apply rigorous dependency analysis in order to construct data representation of the entities and relationships. To apply mathematical dependency analysis without understanding and articulating the underlying policy can be compared to the experience of shooting in the dark.

Information modeling is a "top-down" procedure rather than "bottom-up;" this means that it starts with large conceptual chunks and decomposes them into smaller pieces (ultimately, data elements). Design procedures that start with dependency analysis are mired down with voluminous detail right at the start of the process. Practical experience has shown this to be unworkable in a surprising majority of business applications.

Major problems in modern system analysis efforts are the volume of detail and the time-consuming nature of the analytical process itself. Information modeling eases these problems significantly by concentrating effort on a top-level view of the real-world system being studied. In this monograph, sophisticated machinery is developed to allow the model designer to quickly seek and identify the major components of the model, and to verify their validity by policy definition procedures. Policy analysis can be conducted efficiently, and system models produced quickly and effectively. Mathematical dependency analysis can be separated from the essential modeling process, so that users and managers may quickly grasp the meaning of the model without an arduous foray into the theoretical world of normal forms and relational data representations.

Entity-relationship modeling has been used in management research, general system analysis, and data processing for more than twenty-five years. Its application to database design began in 1976, when Professor Peter Chen of UCLA put forth the entity-relationship approach to system analysis and modeling as a unified view of database design. Since then, E-R modeling has gained widespread popularity and is quickly becoming the foremost approach to the design of databases.

However, the approach is still relatively intuitive with regard to the conceptual building blocks of the model. Practitioners of the approach have had to face difficult design problems due to the generality and ambiguity of the design constructs. Questions such as the following indicate the extent of the problems:

"How much of my real-world system is a single object?"
"Is CONTRACT a single object or is it a class of objects?"
"Do I have one relationship between CONTRACT, CUSTOMER, and PRODUCTS, or do I have two relationships?"
"Where do I start the E-R model, and how do I know when I'm done?"

Questions like these are common among system analysts who are struggling to master the black art of entity-relationship modeling. To answer these questions, *Fundamental Concepts of Information Modeling* sets forth a new set of definitions of the key concepts: objects, relationships, and unified models. The new definitions are intended to remove most of the mystery about these key concepts by eliminating their extraordinary generality and ambiguity.

Objects are treated first. They are defined in terms of their functional interaction between the real-world system and the observer (user or analyst). The vague notion of "object" is split into distinct "object types" on the basis of the multiple roles that an object may sustain in the real-world system. The process of abstraction is given a new, and substantially deeper, definition within the modeling process.

Relationships are treated next in full generality. They too are rigorously defined so that a generalized relationship between N objects does not become a muddle of separate relationships thrown together helter-skelter.

The book puts great emphasis on rigorous methods of *defining* both objects and relationships with precision. Definition procedures are treated in detail. Finally, concepts are introduced for measuring the quality of the entire model.

Fundamental Concepts of Information Modeling provides the database analyst/designer with the following information:

- full definitions of key semantic concepts of information modeling
- explanations of where the semantic concepts are used
- decision rules to guide the analyst or designer in the application of the semantic concepts to the modeling process

- heuristics and modeling strategies to assist the analyst
  or designer in formulating a systematic approach to the
  development of the model

The monograph is divided into three parts: Part One describes the basic problems that confront the model designer and tells why new conceptual machinery is needed to make entity-relationship analysis productive. Part Two develops the new analytical machinery, which is the heart of the book. Part Three ties all of the ideas together and formulates final definitions for the principal semantic concepts. It also shows how the concepts and techniques developed in this book fit into a complete discipline of information modeling.

In addition, *Fundamental Concepts of Information Modeling* contains a Glossary of technical terms and a Bibliography of works that I have referenced throughout the development of my ideas and preparation of this text. I would advise my readers to use the Glossary freely as a reference tool, as precise in-text definitions are not always given for the many new terms.

Finally, I acknowledge the encouragement and helpful criticism of my friend and colleague Andrei Jezierski. Also, I am grateful for the useful insights of Adrian Baer.

I also thank Wendy Eakin of YOURDON Press for her support and intelligent editorial assistance. I am indebted as well to the efforts of Janice Wormington and Lorie Mayorga, also of YOURDON Press, for their contributions during the editing and production processes.

April 1981

Matt Flavin
New York City

Part One
The Problem

Part One describes the basic problems encountered when the analyst tries to model a complex problem domain. In Chapter 1, information modeling is shown to represent a total approach to the task of system analysis. The logical components of an information model are explained and illustrated, and the methods of information modeling are summarized in outline form.

Chapter 2 addresses the problem of applying modeling concepts to the problem domain. The difficulties of the intuitive approach are discussed. In addition, the problems of intuitive analysis are shown to be resolved by the formulation of highly disciplined modeling concepts, whose logical foundation will be developed in later chapters.

1

Practical aspects of database design

1.1 ■ Databases describe real-world systems

Databases keep information that is needed to support manual or automated processes in real-world systems. These systems usually are business functions — such as marketing, manufacturing, or accounting — found in business or corporate enterprises. Information kept in databases typically describes the state of some business entity, event, or transaction. A complete database describes all business entities, events, and transactions that are associated with a given business function.

The logical design of a database must reflect the structure of the business system it describes. It must be a *model* of that system, not of the way a system works, or the way that inputs are processed to become outputs. Databases model the *state* of the business system at any point in time. The state of a system is usually specified by showing the relationships that hold between the components of the system at any point in time. For example, if a service company writes a contract with a client company to complete a series of projects for that client company, then the business transaction of signing the contract establishes legal relationships between the service company and the client company involving service projects to be rendered.

Figure 1.1 is a diagram of the example given above. The business entities are represented by rectangular boxes. The operation (signing the contract) is represented by the ellipse. The resulting relationship between CONTRACT, SERVICE COMPANY, CLIENT COMPANY, and SERVICE PROJECTS is represented by a diamond. The relative ratios between occurrences of each participating entity are given by the numbers that accompany the relationship lines. The letters N and K signify "possibly more than one." For each CONTRACT, one or more SERVICE PROJECTS will be delivered to the CLIENT COMPANY. Figure 1.1 is called an *augmented entity-relationship diagram*.

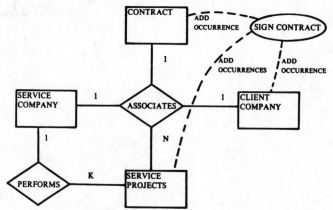

Figure 1.1. Contract-signing operation changes state of database.

This model of a business system can directly serve as a logical database design. The model is very simple; it is constructed of three components: entities, relationships, and operations. These simple components can be used to model business systems of any complexity.

Each of these components must be definable in terms of the business policy that governs the business system. Policy research and analysis are therefore basic to the modeling process. Entities and relationships are described by unit facts, or data elements. Data elements are another component of databases, although they are not shown on Fig. 1.1. The data elements describing CONTRACT might be

- CONTRACT NUMBER
- DATE SIGNED (CLOSED)
- SALESPERSON NAME
- CONTRACTOR NAME
- CONTRACTEE NAME
- PRICE OF CONTRACT
- NAMES OF ASSOCIATED SERVICE PROJECTS

We arrive at the conclusion that logical database design is thoroughly dependent on an analysis and modeling of the business system that the database is intended to describe. Information modeling pays great attention to the way that business systems are analyzed and modeled. The next section will put the problem of analysis into deeper perspective.

1.2 ■ Practical techniques of analysis and modeling

Current practices of analyzing business systems for the construction of databases are largely *intuitive* and *undisciplined*. In general, most database practitioners know that the logical design of a database will involve the representation of business entities and business relationships. The search for these entities and relationships is a primary objective in practical database design. But the process is largely undisciplined, and most database designers encounter significant difficulties when trying to apply intuitive analysis procedures to complex business systems.

Let me give an example of the problems of intuitive analysis. In the banking industry, it is often important to know how a given customer measures up as a business risk. In order to do this, it is necessary to review that customer's complete standing with the bank. However, most corporate customers have multiple business relationships with the bank: They may be depositors, recipients of loans, or users of bank services such as cash management or trust management. Traditionally, each of these distinct relationships was handled by a separate file for that customer, and there would be as many files as there were business relationships between that customer and the bank. However, in an integrated database, all customer information must be present in some form. For the logical database designer, the questions are: How is a customer to be represented? Is a given customer equivalent to one business entity, or multiple business entities, one for each business relationship? How does one unify all the data describing a bank customer, but still provide modularity in the design that guarantees the stability of logical data structures?

This very real problem for database designers is caused by a lack of precision and agreement upon what is meant by the term "business entity." This problem of customer representation can be solved with a set of rules and definitions that remain consistent and universally applicable across all such problem domains. I develop the solution in Part Two of this monograph.

The concepts of entities and relationships are intuitively very appealing and of great heuristic value in motivating the analysis of complex business systems. But as intuitive concepts, they are difficult to formalize and nearly impossible to reduce to a mathematical format.

The great power of these concepts in organizing a body of knowledge is also their intrinsic limitation whenever we attempt to apply them with precision, as in the example above.

Some authorities have argued that entity-relationship analysis is so general and so imprecise that it should be thrown out and replaced by something with greater mathematical precision. Certainly, such a view is a point of great contention. These authorities have taken their cue from the successful ideas of Codd of IBM. Codd proposed a method of representing data in the form of two-dimensional tables, called *mathematical relations*. In order to ensure that data stored in tables could be retrieved accurately, Codd augmented relational theory with the theory of *functional dependencies*, which guaranteed that all the data in a table could be accurately retrieved if *functionally dependent* on the key of the table. Functional dependence is a relationship between two data elements that guarantees that one of the data elements will determine the content of the other data element. The theoreticians maintain that databases do not contain entities and relationships, as such, but rather that databases are composed of mathematical relations (tables) that have been assembled from an exhaustive process of compiling functional dependencies.

Before I launch into a critique of this view, I would remind you that the vast majority of practitioners are making do with their intuitive methods. Although the intuitive method is technically flawed, seasoned professionals find it immeasurably preferable to an excursion into mathematical theory. As a trained mathematician, I fully appreciate the plight of the practitioner who has to deliver results in a timely fashion in order to meet business deadlines. There *is* a middle ground open to the database practitioner between the undisciplined methods of intuitive analysis and the overly detailed procedures of mathematical dependency analysis. This middle ground is called information modeling, and I shall further develop the concepts of this discipline in Part Two of this monograph.

Analysis is a process of imposing structure on an unstructured problem domain. The structure disposes the problem solver to create workable solutions to the problem at hand. The structuring of the problem domain facilitates the problem-solving process. The resulting structure is a model of the problem.

True analysis always begins with an unknown, that is, the *unstructured problem domain*. It is the job of the analyst to research the facts, exercise critical judgment, and impose order and structure where none existed. Analysis techniques can be refined and the tools can be honed, but there is no way to completely eliminate uncertainty from the analytical process.

Unjustified faith in the power of mathematics has led some authorities to conclude that all uncertainty and human judgment can be eliminated from the analytical process. The need to refine our techniques has been confused with the total elimination of uncertainty altogether. This is a fundamental fallacy.

In logical database design, there can be no shortcut around the basic requirement to research, define, and structure business policy. Undisciplined entity-relationship analysis is flawed by imprecision of technique, but it retains the basic framework of business policy analysis. Entity-relationship analysis, with its intuitive motivation, is fitted to business policy analysis at the proper level of abstraction for framing and defining business policy. Business policy is the bedrock against which logical databases are validated. A logical database design procedure *must* incorporate a means to validate the logical database design against stated, and verified, business policy.

Information modeling provides a *disciplined* procedure for carrying out an entity-relationship analysis of a business problem domain, which can be validated against business policy. It does not make idle promises to eliminate uncertainty or critical judgment from the analytical process. It does not promise to eliminate thinking from analysis and modeling. It makes no claims that critical problem-solving efforts can be transferred to a computer. Information modeling *does* enhance the accuracy, clarity, and speed of the analytical process. The new concepts of *well-defined* object type, *well-defined* relationship, and *well-formed* model eliminate confusion and fuzziness in the modeling process.

The disciplined techniques of information modeling optimize the precision of a top-down analytical procedure. The formulation of functional dependencies is nearly a triviality given the logical framework of an entity-relationship model; but to start with the formulation of functional dependencies between data elements and then derive a valid entity-relationship model is a practical impossibility.

The message of information modeling is quite simple: The tough job of analyzing a business problem domain has always been ignored and treated as a black art. We can never reduce the intellectual work of analysis to a computer program, but we can develop better tools and discipline our techniques so as to take a large measure of the mystery out of this black art. Information modeling provides the means.

1.3 ■ Components of information models

The five logical components of an information model are object types, relationships, operations, data elements, and regulations, and are described on the next page.

O   OBJECT TYPES: These are the business entities in a business model. Examples are persons, places, things, documents, organizations, agreements, or policies. Object types play a specific role in the system being modeled.

O   RELATIONSHIPS: Relationships are named associations between two or more object types. They are the result of interactions between the participating object types, or logical associations between the object types.

O   OPERATIONS: An operation is an action that changes the state of the system being modeled, and thus the information model of that system. Operations can be identified with business transactions and events.

There are two classes of operations: *standard* and *user defined*. Standard operations are listed below.

ADD
DELETE } component type

ADD
DELETE } component occurrence

ENTER
REMOVE } subcomponents

MODIFY
REWRITE } contents

User-defined operations are defined in terms of standard operations. SIGN CONTRACT is an example of a user-defined operation.

O   DATA ELEMENTS: Data elements are unit facts (for example, PHONE NUMBER) that describe object types or relationships. Data elements are associated with object types and relationships if they describe those components in a meaningful way.

O   REGULATIONS: A regulation is a rule that governs the content, structure, integrity, and operational activity of the model. It applies to the model as a whole. In general, regulations express high-level policy constraints.

Each logical component is described by six subcomponents: name, definition, data content, data structure, allowable operations, and data dependencies. These descriptive subcomponents are defined below.

O   NAME: Each logical component has a unique name within the information model. The name may consist of multiple words.

○   DEFINITION: Each component is defined in terms of the role that it plays in the model. Business policy is generally the source of valid component definitions.

○   DATA CONTENT: Object types and relationships are associated with data elements that describe them in some way. The set of associated data elements defines the data content of object types or relationships.

○   DATA STRUCTURE: Data elements associated with some component can be classified as single-valued or multi-valued for that component. For example, if the object type is EMPLOYEE, then the data elements NAME and SOCIAL SECURITY NUMBER have only one value for each EMPLOYEE. But the data element CHILD NAME may have multiple values for any EMPLOYEE. Single-valued data elements are grouped in a structure called the "base segment." Multi-valued data elements are grouped into one or more "dependent segments." The collection of all segments makes up the data structure associated with the given component.

○   ALLOWABLE OPERATIONS: All operations that can be performed on a given component are described by "allowable operations." Since operations change the state of the model, each allowable operation is described by a set of pre- and post-conditions. Pre-conditions are the conditions that must be true for the model if the operation is to be performed. Post-conditions are the conditions that must be true for the model immediately after the operation has been performed.

○   DATA DEPENDENCIES: A data dependency is a rule that states that if some condition A is true, then some condition B must be true. Data dependencies are the chief tool for logical integration of an information model. An example of this is: If SALARY GRADE = 13, then SALARY > 38000. Data dependencies are associated with the component to which they apply.

The following sections further describe the five logical components; their examples are based on Fig. 1.1. Five Exhibits give examples of documentation for each component, and are written in a policy-documentation format called Formal English.

1.3.1 ○ Object types

Object types are the fundamental components of an information model, and correspond to the active components of the system being modeled. They are the business entities of the business system. Four object types are depicted in Fig. 1.1 on page 4.

The underlying theory for the identification of object types in an information model will be developed in Part Two of this monograph. Exhibit 1, shown on the following pages, is an example of a formal definition of an object type.* Note that the description is composed of the six subcomponents mentioned above. You will notice that the policy DEFINITION for CONTRACT is quite extensive. This is because object types are generally the focal points of the system being modeled. Usually, object types are defined by the business policy that governs the system being modeled.

Object types and relationships are usually described by way of data elements, which are attributed to the object types as properties or attributes. These data elements are listed in Exhibit 1 in the DATA CONTENT subcomponent. Data elements are associated in groups called *segments*. Segments determine the DATA STRUCTURE of object types.

Each component of a model is capable of being operated upon. Object types can be added to or deleted from a model. Since each object type (for example, CLIENT COMPANY) defines a class of individual occurrences, it is possible to add occurrences to or delete occurrences from any given object type. I want to distinguish between *types* and *occurrences:* To delete an occurrence of an object type does *not* affect all remaining occurrences. But to delete an object *type* would result in the deletion of *all* occurrences of that object type from the model.

The subcomponent ALLOWABLE OPERATIONS defines which types of operation can be performed on the model component. You will also notice that each allowable operation is associated with pre-conditions and post-conditions. The pre- and post-conditions define the transition of states in the information model when an operation is performed. *Thus, an information model combines entity-relationship analysis with state-transition analysis.* The final subcomponent is the description of DATA DEPENDENCIES, which are the chief tool for the logical integration of an information model or database.

Since object types are the most important logical components of an information model, their description tends to be lengthy. Exhibit 1 is proof of this.

---

*Two conventions of Formal English occur in the Exhibits and should be noted: First, the caret serves as a pointer to policy statements. Second, the period used between object types or relationships and data elements in the DATA DEPENDENCIES subcomponent indicates that the word or words to the left of the period function as a qualifier or owner of the data element immediately following the period.

EXHIBIT 1

COMPONENT: OBJECT TYPE
NAME: CONTRACT
DEFINITION:

PURPOSE:
> A contract is a legal agreement between the service company and the client company.
> A contract defines one or more units of work called service projects.
> For each service project, a contract defines a set of project deliverables.
> A contract is the unit of sale of project services.

PROPERTIES:
> A contract has a begin date and an end date.
> A contract always defines a final set of deliverables.
> All contracts have a fixed cost.
> Contract costs are settled at the time of signing the contract.
> Each contract has a unique contract number.

SYSTEM EFFECTS:
> A contract can be renegotiated before or after its end date.
> A contract can be terminated while in progress for sufficient cause.
> A contract can be nullified any time within the first 10 days after it has been signed.

OBJECT EFFECTS:
> A contract in progress may impact the ability of the service company to commit resources to other clients.

ASSOCIATIONS:
> A contract defines a relationship between the client company, a service company, and service projects.
> Multiple contracts between the service company and a client company may be in force at any time.

DATA CONTENT:
> contract number, single-valued
> date signed, single-valued
> salesperson name, multi-valued
> salesperson number, multi-valued
> contractor name, single-valued
> client name, single-valued
> client address, single-valued
> client phone, single-valued
> contract cost, single-valued
> project number, multi-valued
> project name, multi-valued

DATA STRUCTURE:
> base segment: contract
unique identifier: contract number
attributes: contract number
date signed
contractor name
client name
client address
client phone
contract cost

> *dependent segment: salespersons*
> *unique identifier: salesperson number*
> *attributes: salesperson number*
> *salesperson name*
> *dependent segment: projects*
> *unique identifier: project number*
> *attributes: project number*
> *project name*

ALLOWABLE
OPERATIONS:

(1) Add occurrence
PRE-CONDITIONS:
> *For each contract, there is a client company such that the clien*
> *company is defined before the contract is entered into the mode*
POST-CONDITIONS:
> *For each contract, there are service projects such that the servi*
> *projects are defined in the model if they are named in the contr*
(2) Delete type
POST-CONDITIONS:
> *For each service project, service projects are deleted from the n*
> *For each declared relationship, the definition of the relationshi*
> *is modified;*
> *or*
> *the relationship type is deleted*
> *if contract is a participating object type in the relationship.*
(3) Delete occurrence
PRE-CONDITIONS:
> *For each contract, there is a client company, a service compan*
> *and service projects such that the contract associates*
> *service projects with a client company and the service company.*
> *For each deleted contract, there are service projects such that*
> *either*
> *the service projects are deleted from the model;*
> *or*
> *the service projects are associated with the client company*
> *via some new contract.*
(4) Sign contract

DATA
DEPENDENCIES:

> *For each contract, there is a service company,*
> *there is a client company, and there are service*
> *projects such that contract.contractor name =*
> *service company.name*
> *and*
> *contract.client name =*
> *client company.name*
> *and*
> *contract.project number =*
> *project.number.*

1.3.2 ○ Relationships

Exhibit 2 is the description of the second logical component, relationships. Relationships also possess a unique NAME and policy DEFINITION. *A relationship is an association between object types that is the result of interactions between object types, or logical dependencies between object types.* In general, most relationships are the result of business-defined interactions between object types.

You will notice that relationships have data elements attributed to them. Therefore, they are described by DATA CONTENT and DATA STRUCTURE. The reason for this is simple: Object types are associated by matching their unique identifiers; combinations of the identifiers determine *occurrences* of the relationship.* Therefore, the attributes of a relationship are the unique identifiers of all object types that participate in the relationship. Notice that relationships are also described by ALLOWABLE OPERATIONS and associated DATA DEPENDENCIES.

1.3.3 ○ Operations

Exhibit 3 is an example of the description of user-defined operations. Operations, as previously defined, are actions that change the state of the model.

A user-defined operation is a macro or composite operation that is composed of a combination of standard operations. In Exhibit 3, SIGN CONTRACT is a composite of various ADD operations.

Since user-defined operations are composites of standard operations, the DATA DEPENDENCIES are described by complex pre- and post-conditions.

1.3.4 ○ Data elements

Exhibit 4 is an example of the fourth logical component, data element. A data element is the smallest unit of information in an information model or database. Data elements do not stand alone in a database. They are always associated with (attributed to) object types or relationships.

The DATA CONTENT subcomponent describes the range of values that can be assumed by the data element. The DATA STRUCTURE subcomponent describes the type and format of the data element. Even though data elements do not stand alone in an information model, they are subject to ALLOWABLE OPERATIONS and DATA DEPENDENCIES wherever they are attributed within the model.

---

*Note that a *unique identifier* (primary key) is a data element that contains a unique value for each distinct occurrence of the object type.

EXHIBIT 2

COMPONENT: RELATIONSHIP
NAME: PERFORMS
DEFINITION:

    BASIS:
> For the service company, there are service projects such that the service company performs the service projects if the service projects are contracted to some client company and the service projects are not performed by a subcontractor.

DATA
CONTENT:
> service company name, single-valued
> client company name, single-valued
> contract number, single-valued
> project number, single-valued

DATA
STRUCTURE:
> base segment: performs
> unique identifier: project number
> attributes: project number
>   service company name
>   client company name
>   contract number

ALLOWABLE
OPERATIONS:

(1) Add occurrences
    PRE-CONDITIONS:
> For each occurrence of performs, there is a service company and there is some project such that the service company and project must exist in the model before the associating occurrence of performs is added.
    POST-CONDITIONS:
> For each occurrence of performs, there is a service company and there is some project such that the service company and project exist as long as the associating occurrence of performs exists.

(2) Delete occurrences
(3) Add type
(4) Delete type

DATA
DEPENDENCIES:
> For each occurrence of performs, there is a project, service company, client company, and contract such that performs.service company name = service company.name
> and
> performs.project number = project.number
> and
> performs.client company name = client company.name
> and
> performs.contract number = contract.number.

EXHIBIT 3

COMPONENT: OPERATION
NAME: SIGN CONTRACT
DEFINITION:

EFFECT:	> For some contract, there is a client company, service company, and service projects such that sign contract establishes a contract between the service company and the client company for rendering service projects.
ACTIONS:	> For each occurrence of sign contract: add contract occurrence, and add service project occurrences, and add client company occurrence if it does not exist in the model.

ALLOWABLE
OPERATIONS:

(1) Add type
(2) Delete type
(3) Modify type

DATA
DEPENDENCIES:

PRE-CONDITIONS:
> The service company must exist in the model.
POST-CONDITIONS:
> For each occurrence of sign contract:
the model will contain an additional
occurrence of contract;
and
the model will contain additional
occurrences of service projects;
and
the model may contain an additional
occurrence of client company.
> For each contract, there is a client company,
service company, and service projects such that
contract.contractor name =
service company.name
and
contract.client name =
client company.name
and
contract.project number =
service project.number.

EXHIBIT 4

COMPONENT: *DATA ELEMENT*
NAME: *CONTRACT NUMBER*
DEFINITION:
> *A contract number is a unique identifier for each unique contract.*
> *A contract number is composed of a customer number and a contract sequence number.*
> *A contract sequence number is assigned in chronological sequence.*

DATA
CONTENT:
> *Value range of the customer number is 0 to 9999.*
> *Value range of the contract sequence number is 0 to 9999.*

DATA
STRUCTURE:
> *Type is decimal with no decimal places.*
> *Format is XXXXYYYY where XXXX is the customer number and YYYY is the contract sequence number.*

ALLOWABLE
OPERATIONS:
(1) Delete type
POST-CONDITIONS:
> *There exists an element X such that there is a unique value of X for each distinct contract.*
(2) Add occurrence
PRE-CONDITIONS:
> *There exists a customer such that the customer number component of the new contract number is the unique identifier of that customer.*
> *The new occurrence of contract number is distinct from all contract numbers existing in the model.*
(3) Delete occurrence.
PRE-CONDITIONS:
> *There exists some contract such that the contract number to be deleted is the unique identifier of that contract.*
POST-CONDITIONS:
> *See object type = contract, allowable operation = delete occurrence*

DATA
DEPENDENCIES:
> *For each contract number, there is a customer such that contract number.customer number = customer.customer number.*

1.3.5 ○ Regulations

Exhibit 5 shows the fifth component of an information model: regulations. *A regulation is a rule that applies to the entire model.* It possesses a NAME and a policy DEFINITION. Like operations, it contains no DATA CONTENT or DATA STRUCTURE.

Regulations are subject to standard operations and generally result in complex DATA DEPENDENCIES.

1.4 ■ Information modeling as a whole

Figure 1.1 was a *graphic* specification of an information model. Exhibits 1 through 5 are *analytic* specifications of an information model. Now we may ask: What is the relationship of the information model to a database?

DEFINITION: **database**

a systematic and logically organized collection of facts. The logical structure of the data is derived from an information model. A database will store all occurrences of object types, relationships, and data elements.

Traditionally, logical models have served as logical designs for databases, but they have never been regarded as part of the total database system. Information modeling takes a different view: An information model is composed of the component specifications *and* a database (see Fig. 1.2).

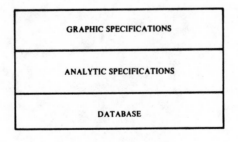

GRAPHIC SPECIFICATIONS

ANALYTIC SPECIFICATIONS

DATABASE

INFORMATION MODEL

Figure 1.2. Three partitions of an information model.

EXHIBIT 5

COMPONENT: REGULATION
NAME: PROPERTY INHERITANCE RULE
DEFINITION: > For each object type X, the attributes
                   of object type Y are also attributes of X
                   if X is the subtype of Y.

ALLOWABLE
OPERATIONS: (1) Delete type
                 (2) Modify content

DATA
DEPENDENCIES: > For each element X, there are object
                   types Y and Z such that X is removed from
                   Y if Y is a subtype of Z and X is removed from Z.

Therefore, an information model is a total system composed of policy-defined conceptual structures *and* data structures that represent each conceptual structure with actual data values.

An information model is also *dynamic.* It can specify changes of state in the system being modeled. These state-changes are described by the pre- and post-conditions associated with each operation. Thus, the dynamics of an information model are state-oriented rather than procedure-oriented. Some forms of system analysis are strongly procedure-oriented. In them, state-changes are specified by sequences of actions, which are called procedures, processes, or functions.

Information modeling is more abstract in that it specifies before and after states of an operation without specifying the exact sequence of actions that implement an operation. Information modeling is the union of two types of system analysis: entity-relationship analysis, which defines the *subject matter* of the system being modeled; and state-transition analysis, which defines the *dynamics* of state-changes in the system being modeled. The information modeling process can be split into two major phases: *analysis* and *representation* (see Fig. 1.3).

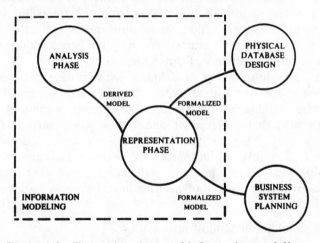

Figure 1.3. The major phases of information modeling.

The analysis phase is composed of procedures for the analysis of the business problem domain. This phase identifies the object types, relationships, operations, and data elements that compose the model of the business domain. Also, augmented entity-relationship diagrams are produced, the analytic specifications for all model components are partially completed, and policy definitions are created in this phase.

The representation phase is concerned with creating data structures for the components that are represented in the database, and completing the logical integration of the model. Data dependencies, pre- and post-conditions, and regulations are specified in this phase.

1.5 ▪ The domain of information

Before we begin building information models, it is necessary to specify which real-world system, or which part of it, is the subject of our model.

DEFINITION: **domain of information**

a real-world system, or a bounded portion of that system, that is the subject of an information model. It is represented in an information model by semantic components (objects, relationships, and so on), data structures, and data elements.

The information model for a given domain of information will be shared by a community of users who have an interest in the given domain. For example, a domain of information called "marketing operations" would be of interest to the members of the marketing department and to members of top management.

Whenever information modeling is used to design databases, the information model for a given domain may be implemented by one or more physical databases. This means that for each domain of information, there will usually correspond one or more operational databases.

We can ask, "How do you specify the contents of a domain of information?" The information model *itself* is the specification of content for a domain of information. However, there are degrees of specification for the domain of information. Levels of specification are listed below:

- name of domain of information

- entity-relationship diagram of domain

- E-R diagram and complete set of required data elements for the domain

- fully defined and completed information model of the domain of information

1.6 ■ Importance of policy research

In order to complete our information model, it is necessary to define each component of the model (objects, relationships, data elements, and operations) unambiguously and consistently in the context of their meaning in the real-world system. This is not an easy task: Producing meaningful definitions of model components that are agreeable to a diverse user community can be a challenging and exhausting task. In extreme cases, it may establish limits on the model-building itself. In general, however, it can be accomplished on the basis of persistence and a clear conception of what the model builder is trying to accomplish.

It is important to note that an information model builder has to be part sleuth in order to construct a successful model or database. An information model cannot be created from the imagination; the designer must go to the user community as the ultimate source of knowledge about the domain of information.

Definitions of object types, relationships, and the other model components will depend on discovering the *rules, laws, and conventions* that govern the composition, structure, and operation of the system being modeled. These rules and laws are the *policy* of the system being modeled. Most often, policy is not written down neatly in a manual; it must be extracted piece by piece from the combined knowledge of members of the user community. Production of an information model is squarely based on a policy research and analysis of the system being modeled. Consequently, a completed information model is a *specification of known policy* for a given real-world system.

1.7 ■ Requirements analysis and the data dictionary

In addition to policy research, which is used to define the model components, the model designer needs to know which data are needed to satisfy users' requirements for information describing the model. Users will request that each component be described by some set of data elements. Specification of the data element content of an information model will depend on an analysis of user requirements. Policy research and requirements analysis are thus concomitant activities in the process of model derivation.

Before the set of data elements is fully integrated into an information model (assigned as attributes to objects and relationships), it must be collected into a document called the *data dictionary*. The data dictionary is usually organized as an alphabetized listing of data elements. For each data element (e.g., BIRTHDATE, PHONE NUMBER, SALARY), there is an entry in the data dictionary. Within each entry, there is a set of descriptions of the data element, including

- formal name of data element
- aliases of data element
- definition
- data type specification
- data format specification
- content specification
- integrity constraints

As a rule, compilation of a data dictionary for a domain of information is easier and less problematical than construction of the full information model. The reason is that objects and relationships define organizational policy at a higher and more sensitive level. For this reason, compilation of the data dictionary is one of the first steps to be completed in the model-building process.

1.8 ▪ Model semantics, analytical procedures, specification policy

Information modeling is a complete and self-contained methodology for the construction of a high-level business model. All of the difficult stumbling blocks associated with the task of analyzing a complex business domain are eliminated through the introduction of carefully designed methods and procedures. Three fundamental problems are addressed in the questions and answers that follow:

Q. How can I apply the concepts "object" and "relationship" to the problem of modeling a complex business domain with accuracy and consistency?

A. Each of these intuitive concepts can be associated with a strict set of rules that govern the way in which they are applied to the problem domain. This set of rules must be obeyed each time the object or relationship concept is used in an information model. By satisfying the set of rules, the analyst can be assured that he or she is applying the concepts accurately. Since the same set of rules is invoked for each application of these concepts, the analyst is assured of consistent application of these concepts. These sets of rules determine the semantics, or meaning, of each modeling concept. Part Two of this monograph will develop the sets of rules needed to support each modeling concept. The rule-bound concepts of the well-defined relationship, the well-defined object type, and the well-formed model are formulated in Part Three.

Q. What methods can I use to correctly and completely identify the relevant object types, relationships, and operations that belong in an information model of my business problem domain?

A. Search procedures developed in Part Two can be used to assist the analyst in viewing the problem domain in a productive way and in asking the most pertinent questions about the structural features of the problem domain. These analytical procedures can provide important heuristics for motivating the process of discovering and identifying the relevant structural components. Information modeling makes use of five fundamental analytical procedures: functional analysis, scenario analysis, transaction analysis, abstraction analysis, and anchor-point analysis. These procedures are fully explained in Part Two.

Q. How can I write policy specifications of model components in descriptive English so that the intent of the policy is clear even when the problem structure is complex?

A. The English language can be constrained to follow a set of logical and structural rules that guarantees that each policy statement expresses its intended policy in a clear and unencumbered way. These rules govern the way policy statements can apply to complex combinations of model components. Once mastered, the rules form a uniform and straightforward way to express policy of any degree of complexity. These rules form a discipline called Formal English.

2

Definition of the problem

Now that we have discussed in general terms why information modeling is necessary, what its components are, and how it is used, we can turn to the specific questions that will be raised in this chapter and more fully answered by the definitions and rules of Part Two.

For the beginning database designer who faces building an information model for the first time, there are three basic questions:

- What kinds of semantic structures am I looking for?
- How do I find them?
- How do I describe them unambiguously?

Database design can be difficult because it forces the designer to transform private, intuitive knowledge into an accurate public model of a system. The difficulty (and challenge) lies in trying to convert vague, unstructured intuition into something that is structured and precise.

This is never an easy task, but we can reduce the difficulty by narrowing the scope of design possibilities and providing targets for the designer. Let me summarize the major problems of semantic information modeling in the following three sections.

2.1 ■ Ambiguity and generality of the object construct

An object can be a person, place, thing, event, policy, procedure, agreement, or concept, among other things; in general, it is anything that can be named. As I define them, objects range from very concrete items, such as "your shoes," to very abstract entities, such as the "quality of mercy." Often the range of abstraction open to a database designer, or system analyst, is enough to cause indecision and confusion. For example, the object known as "insurance contract" may be applied to all individual insurance contracts written under all insurance plans. But suppose that the policy governing the contract is distinct for each insurance plan. We face the question of whether the object "insurance contract" has meaning in our real-world system, or whether it should be replaced by a set of specific types of insurance contracts, one for each plan.

Another difficulty of the object construct may be seen in the example of a freight shipment. Suppose that "freight shipment" roughly means that something is moved from point A to point B. In real-world freight systems, trying to define shipment is where the fun begins. First, the contents of a shipment are almost never describable in a uniform fashion. Decisions about which containers will carry the contents of a shipment are made at the time of the shipment. During the progress of the shipment, the contents may be split and carried separately, only to be reunited at the destination.

There may be multiple destinations, a portion of the contents being delivered at each point; or the contents may change form along the way. For example, iron ore can begin its itinerary at a mine, then be carried to a steel refinery, and then arrive at its destination as girders. Needless to say, the object "freight shipment" poses quite a problem to the system analyst.

The notion of "marriage" introduces a third type of problem. In some situations, marriage is a relationship between two types of objects, males and females. In other situations, it is treated as a legal entity that is recognized by various governments.

These three examples of "insurance contract," "freight shipment," and "marriage" show that a great deal of fuzziness can surround the object construct when it is used casually or informally. Let us now construct a working definition for what we mean by *fuzzy object*.

DEFINITION: **fuzzy object**

an object is called fuzzy if it is *not* possible to define its role or function in the system being modeled, identify unique occurrences of the object, distinguish it from other types of objects in the model, distinguish it from other semantic constructs in the model (for example, relationships), or define its characteristics in the same way for all occurrences.

Because of the intuitive and fuzzy character of the object construct, some authorities* have argued that the notion should be thrown out of rigorous information models altogether. They have advocated that the object construct be replaced by something more mathematically rigorous (e.g., mathematical relations). However, all proposed mathematical replacements have required extensive buttressing in order to capture the original meaning of the information model.

It is my contention that each semantic construct (object, property, relationship) can be made meaningful and precise through a careful process of building the logical basis of each construct. This is accomplished by making sure that *each semantic construct conforms in its definition to a set of logical constraints that will guarantee its integrity within an information model.* A preliminary definition[†] of well-defined object type follows.

DEFINITION (PRELIMINARY): **well-defined object type**

a type or class of objects is called well defined if it is possible to precisely define its role or function in the system being modeled, uniquely identify each distinct occurrence, distinguish it from other object types in the model, distinguish it from other semantic constructs in the model, and define its characteristics in the same way for all occurrences.

---

*Codd, Schmid and Swenson, and Smith and Smith are among those authorities who have addressed this concept. References are cited in the Bibliography.
[†]Final definitions are predicated on rules that will be presented gradually during Part Two; when the set of rules is complete, final versions of the definitions for *well-defined object type, well-defined relationship,* and *well-formed model* will be presented.

2.2 ■ Ambiguity and generality of the relationship construct

A relationship is an association between object types. For example, suppose in some given company that the major part of the business is handled contractually. That is, contracts are used to define the goods and services in a business transaction. Salespeople are responsible for closing those contracts with the customers. We could associate CUSTOM-ERS, CONTRACTS, and SALESPERSONS in several ways. Suppose that multiple salespersons could participate in the sales process for any given customer. Suppose also that only one salesperson is allowed to close a contract, and this is determined by management policy. These constraints permit us to determine several different associations between CUSTOMER, CONTRACTS, and SALESPERSONS depending on the way we use the constraints as conditions for the relationship.

Figure 2.1 shows an entity-relationship diagram of an unspecified relationship called X. X can be interpreted in three different ways. In Fig. 2.1, we do not know how the participating objects are associated.

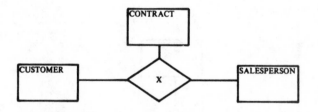

Figure 2.1. An unspecified relationship.*

Figure 2.2 shows that each CUSTOMER is associated with several CON-TRACTs and several SALESPERSONS if at some point in time the given CONTRACTs were closed for that CUSTOMER, and the given SALESPER-SONS participated in some sales effort involving the CUSTOMER at some point. Figure 2.2 does not specify how CONTRACT and SALESPERSON are associated under this relationship.

---

*Note: Figures 2.1 through 2.4 are E-R diagrams. The rectangular boxes represent object types; the diamonds represent relationships; and any number of object types can be connected to a diamond. In Figs. 2.2 through 2.4 are small numbers attached to connecting lines. These numbers show the relative ratios of occurrences of participating object types. The letters M and K signify "possibly more than one."

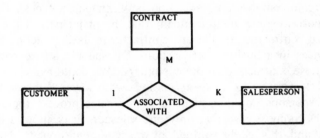

Figure 2.2. An unspecified association between object types.

Figure 2.3 represents the relationship that associates each CUSTOM-ER with a given CONTRACT *and* the SALESPERSONS who participated in the sale of that contract. In the relationship ASSOCIATES, we do know how CONTRACTS and SALESPERSONS are associated with each other.

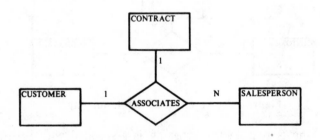

Figure 2.3. An unambiguous association.

The entity-relationship diagram in Fig. 2.4 depicts a relationship that associates each CUSTOMER with a given CONTRACT and the one SALESPERSON who closed that CONTRACT.

In Figs. 2.3 and 2.4, there is no ambiguity in the way that all three object types are associated. However, Fig. 2.1 is completely ambiguous, and is actually a representation of a family of relationships. In this case, the family of relationships is determined by Figs. 2.2 through 2.4.

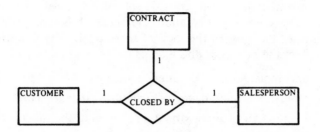

Figure 2.4. A specified relationship showing occurrence ratio.

Just as objects can be fuzzy, so can relationships be fuzzy if we cannot accurately associate the respective occurrences of all participating object types.

DEFINITION: **fuzzy relationship**

a relationship is called *fuzzy* if it is not possible to specify the rules, conventions, or laws that determine the association between participating object types, or specify *unambiguously* how a given occurrence of a participating object type is associated with specific occurrences of other participating object types.

In analogy to the way objects are constrained to retain integrity in the information model, I propose the following constraints on relationships in the form of a preliminary definition for *well-defined relationship*.

DEFINITION (PRELIMINARY): **well-defined relationship**

a relationship between N participating object types is *well defined* if it is possible to specify the rules, laws, and conventions that determine the association between the participating object types; and associate *unambiguously* each occurrence of any participating object type with specific occurrences of other participating object types.

2.3 ■ Need for structural integrity in the information model

An information model consists of object types, relationships, descriptive data elements, operations on those three components, and high-level rules that govern the logic of the model. The following problems can affect the quality of the model:

- incompleteness of declared structures
- unjustified redundancy of structures, data, or both
- unnecessary complexity of structures in the model
- fuzziness of model components
- incompatibility of component definitions within the model

These problems can impair the overall quality of an information model in a way that is similar to the impairment caused by fuzzy objects and fuzzy relationships. The remedy for objects and relationships was to establish a set of rules that govern the application of these modeling concepts. The same can be done for the model as a whole. In Part Two, I shall present the development of the needed rules and procedures. In Part Three, I shall present a final definition of the concept of the *well-formed model*.

Armed with an understanding of well-defined object types and relationships, and well-formed models, the model designer now has a goal from which to build an information model. A high-quality model can be built on the basis of these fundamental concepts without the designer's having to resort to a complex mathematical analysis of the completed model.

Part Two
The Semantics of
Objects and Relationships

Part Two develops the set of rules that are used to define the new modeling concepts, and the analytical procedures that guide the analyst in constructing a model of the problem domain.

Chapter 3 explains the logical underpinnings of the object type concept. Chapter 4 develops the analytical procedures that are used to begin construction of the model. Chapter 5 sets forth the conditions that justify introducing object types into the model. Chapter 6 develops the rules for abstracting new object types. Chapter 7 discusses the quality constraints that apply to the model as a whole. Chapter 8 develops the analytical procedure used to define generalized relationships in an information model.

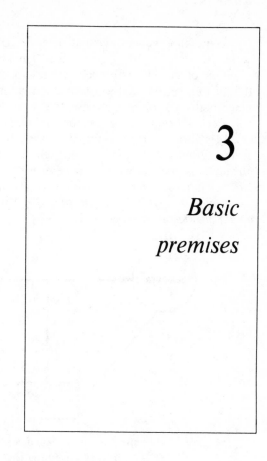

3

Basic

premises

Working from our intuitive concept of object, let us define a series of more detailed premises related to it, which will give us a basis for further developing the rules of information modeling.

3.1 ■ Object types as functions of systems and observers

An object is anything that plays a specific role in the system being modeled; it can be named and described. Most objects are taken from the following categories:

- ○ person
- ○ place
- ○ thing
- ○ event

- ○ policy
- ○ procedure
- ○ agreement
- ○ arrangement

- ○ set of any of the above

Generally, we can say that objects are defined by their properties. There is more to this statement than meets the eye. A general definition of object has eluded writers for thousands of years. I won't attempt to give the final word on this matter; but some progress toward an acceptable definition can be made.

Let us begin with some observations: No object exists in isolation. It is always part of some larger system that is of interest to the observer. Often, the observer himself is part of that larger system. Indeed, it is the observer who determines which properties are of interest. Therefore, *the decomposition of the system of interest into its component objects (entities) is a function of the system, the observer, and their mutual interaction* (see Fig. 3.1).

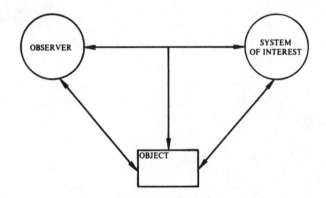

Figure 3.1. Identity of object depends on selection of properties interesting to the observer.

Working definitions of *system* and *object characteristics* will help us to understand this relationship between the object and the observer.

DEFINITION: **system**

a totality that emerges from the net effect of interaction between its elements.

The notion of "object property" can be generalized in the concept of *object characteristic*.

DEFINITION: **object characteristic**

a distinguishing feature that permits an observer to identify an object as belonging to a particular class or type.

There are six kinds of object characteristics, which are defined below.

○ *purpose:* the reason, or set of reasons, for the presence of the object in the system being modeled

○ *defining properties:* the essential, observable features of the object that are associated with all occurrences of the object

○ *object effect on the system:* the observable effect of the action of the object on the system being modeled

○ *system effect on the object:* the observable effect of the action of the system on the object

○ *association with other objects:* the associations abstracted from the object's interaction with other objects or participation in groups

○ *behavior pattern:* the observable pattern of the object's behavior within the system being modeled

Object characteristics are either inherent or assigned by the observer. Inherent characteristics are those characteristics that do not depend on the intervention of the observer; assigned characteristics do depend on observer intervention. Consider the game of chess. The game is defined by the rules of play, which determine the way the chess pieces can be moved around a chessboard. A chess piece is defined by its associated rules of usage, rather than its inherent properties. For example, the queen is defined by its ability to be moved in any direction, for any distance. Any physical object could fill this role, as long as it could be moved about the chessboard. Therefore, a chess queen is a type of object whose characteristics are assigned by the observer (or player). In addition, the physical object filling the role of queen still retains its own inherent characteristics.

DEFINITION: **object function (within the system being modeled)**

a set of characteristics of some object, inherent or assigned, that is of interest to an observer of the system being modeled. The set may contain all or part of the totality of object characteristics.

If the function of an object is preserved as a whole over time, or preserved after normal changes, transformations, or interactions within the system, then we say that the object's function determines an identity of the object over time.

An intriguing, although well-known, property of objects is their ability to carry more than one identity. The example of the chess queen makes this point clearly. If two chess players were stranded without a regulation chess set, they might devise a make-shift set using a bottle cap as a queen. By its inherent characteristics, the cap is a round metallic object used to seal a bottle, but within the game of chess, it is a queen.

This brings home the point that it is the *observer* who is chiefly responsible for the *identification of objects*. Each distinct functional identity of an object determines a different type for that object.

DEFINITION: **object type**

a class of individuals that is characterized by a distinct functional identity, which is preserved as a whole over time in the system being modeled.

The most noteworthy aspect of objects (as part of some phenomenology) is that they form a package of characteristics; that is, *a totality that is capable of entering into external relationships with other totalities.* The fact that objects are unit totalities permits them to have unit identifiers *and* to hide their complex internal structure from external relationships.

For example, consider the nesting of relationships in Fig. 3.2. The boxed relationship ASSIGNMENT relates specific EMPLOYEES to specific PROJECTS. Some of these EMPLOYEE-PROJECT combinations use certain MACHINES within the company. The MACHINES are not assigned to either EMPLOYEES or PROJECTS individually. We have MACHINES related to ASSIGNMENTS, which is the relationship between EMPLOYEE and PROJECT. This nesting leads to complexities in the system model: We find relationships of relationships, properties of properties, and so forth. Characteristics are compounded of characteristics, leading to deeply nested structures.

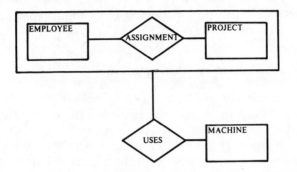

Figure 3.2. Nested structure of a relationship (ASSIGNMENT) within a relationship (USES).

Figure 3.3 resolves the nesting complexity by treating the inner relationship, TASK, as an object. MACHINE is now related directly to TASK, which is named and identified. The inner logical structure of TASK is a relationship between EMPLOYEE and PROJECT. By the mechanism of treating TASK as an object, we have removed the nesting of relationships within relationships, and reduced a complex model to nothing more than objects and relationships with no compounded logical structures.

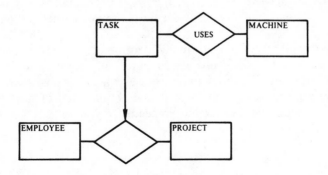

Figure 3.3. Object normalization (TASK) of nested relationships in Fig. 3.2.

Removal of compounded, nested structures enables the creation of a first-order entity-relationship model, having only object-to-object relationships. Object-relationship models can be "normalized"; that is, reduced to objects and first-order relationships only. All compounded, nested structures can be eliminated by converting them to object types in the model. This leads us to a statement of the *principle of object normalization:* Any characteristic of one or more object types that can enter into relationships with other object types will be considered an independent object type in the model.

3.2 ▪ The technique of abstraction

Observers identify different types of objects either by selecting different sets of inherent or assigned characteristics on the basis of their coherence and interest to the observer, or by applying object normalization to compounded characteristics. One might say that objects are identified when an observer *abstracts* from the totality of characteristics in the system being modeled.

DEFINITION: **abstraction**

the derivation of a new object type by the processes of (1) observer selection of object characteristics that are preserved as a whole over time, or (2) object normalization of compounded characteristics.

Observer selection of certain object characteristics implies observer suppression of the remaining characteristics. The set of object characteristics may be taken from the system at large, or may be selected from the characteristics of object types previously defined within the system.

Smith and Smith have formally identified two kinds of abstraction: *generalization* and *aggregation*. It is theoretically possible to identify an indefinite number of abstraction processes that are consistent with the definition above. In Table 3.1, I define these kinds of abstraction as well as two additional abstraction processes: *functional differentiation* and *characterization*.

Table 3.1
Four Modes of Object Type Abstraction

Mode	Explanation	Resulting Object Type
classification (generalization)	treats a class of similar objects as a new object type (see Fig. 3.4)	supertype
association (aggregation)	treats a relationship or association between a set of objects as a new object type (see Fig. 3.5)	associative type
functional differentiation	treats a subfunction of some object type as a new object type (see Fig. 3.6)	subtype
characterization	treats a characteristic of one or more objects as a new object type (see Fig. 3.7)	characteristic type

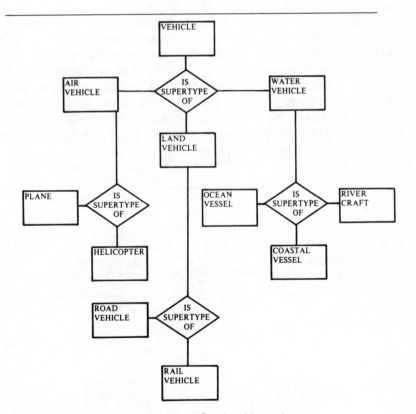

Figure 3.4. A classification hierarchy.

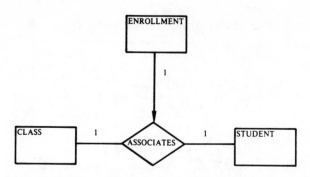

Figure 3.5. An associative abstraction — ENROLLMENT.

Figures 3.4 through 3.7 depict the four defined abstraction processes.

3.3 ■ Deriving object types

Object types are the fundamental logical and structural components of information models. They have a precise identity for the designer and the user. If a single object possesses multiple object types (as in the case of BANK CUSTOMER), we can think of each object type as a distinct, independent "personality" of the object in question. In Fig. 3.6, BANK CUSTOMER has four distinct personalities. Different users of the information model will be interested in different object personalities. This is the prime justification for splitting out distinct object types.

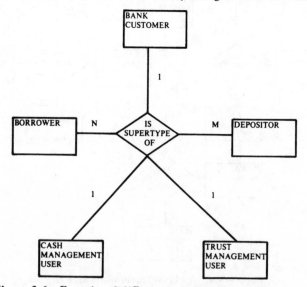

Figure 3.6. Functional differentiation of BANK CUSTOMER.

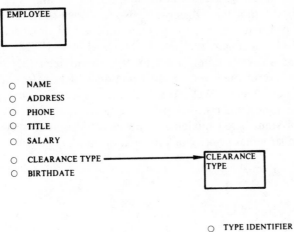

Figure 3.7. Characterization abstraction from EMPLOYEE.

Figure 3.8 is a synopsis of the derivation of object types. The "observer" is either the database user, designer, or both.

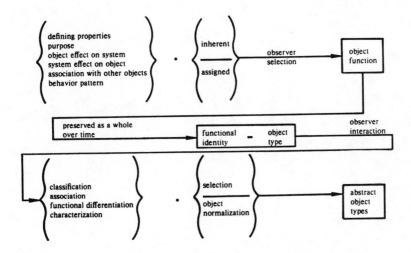

Figure 3.8. Derivation of object types.

In this chapter, we have investigated the logical and conceptual aspects of "object" models. Object models are shown to have a distinctly *psychological* character that is functionally dependent on the interaction between the observer and the system being observed. The functional identity of an object is determined by the phenomena that are of interest to the system observer. Each functional identity specifies a distinct "type" of object. We arrived at the conclusion that the focal points of the modeling process are the object types that play specific roles in the system being modeled. The identification of specific object types is a fundamental priority of information modeling.

4

Search strategies

The basic premises defined in the previous chapter have given us an idea of what components we need to identify in the real world in order to create an information model. In this chapter, we are ready to consider how to find these components.

4.1 ■ The need for search strategies

One of the most difficult problems of any type of system analysis is figuring out how to begin. Where do you start? What are you looking for? How do you find everything?

Traditionally, analysts have had no concrete guidelines in this area. Analytical procedures have been grounded in intuition and general know-how. Information modeling corrects this defect by supplying the analyst with multiple procedures for analyzing the problem domain. These analytical procedures are derived from general conceptual models that are valid for interpreting system phenomena.

Three conceptual models are introduced in this chapter: functional analysis, scenario analysis, and transaction analysis. Each of these three models emphasizes a different method of interpreting a real-world system. Redundancy has been built into the analytical procedures in order to provide the analyst with multiple points of view, which are useful for completeness.

4.2 ■ Functional analysis

Functional analysis emphasizes an input-output model for interpreting systems. The domain of information consists of one or more business functions, by which I mean any coherent set of business activities that serve a common purpose. Marketing, manufacturing, and accounting are three examples of traditional business functions.

Each business function can be split into its component business activities, which perform specific tasks (see Fig. 4.1), and are viewed as input-output procedures.

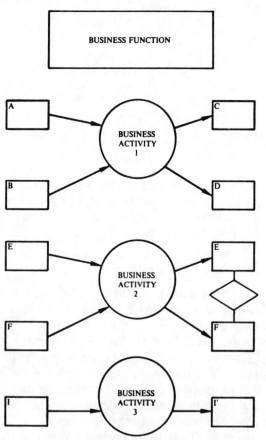

Figure 4.1. Functional analysis.

Business entities may be transformed into other entities, or put into some type of relation to one another, or their internal states may be changed. The business activity is viewed as performing operations on the business object types. By finding the relevant business activities, it is possible to find the relevant object types and relationships. The procedure of functional analysis is presented below.

PROCEDURE: FUNCTIONAL ANALYSIS

FA-1   Identify the domain of information.

FA-2   Identify all business functions that are included in the domain of information.

> In general, several domains of information can be specified for a single business function. Very often, multiple domains of information are specified for a business function depending on the business function's size and complexity.

> *Example:* Manufacturing could be split into domains such as production and materials management.

FOR EACH BUSINESS FUNCTION:

FA-3   Identify the business activities of the business function.

> *Example:* Using an accounting system as an example, expected business activities would be: payroll, accounts payable, accounts receivable, and general ledger update.

FOR EACH BUSINESS ACTIVITY:

FA-4   Identify all object types that are input to the business activity.

> *Example:* Parts are inputs to manufacturing activities.

FA-5   Identify all object types produced by the business activity.

> *Example:* Finished products are outputs from manufacturing activities.

FA-6   Identify all relationships that are established between participating object types by the business activity.

> *Example:* Claims processing would associate claimants with policy holders.

FA-7   Identify all object types that are modified by the activity.

> *Example:* Credit companies update a customer's credit status either to good or unacceptable.

FA-8 Identify all object types or relationships that are used but not changed by the business activity.

> *Example:* Banks would use loan information and credit contract information to process application for credit.

FA-9 Identify all object types that control or implement the business activity.

> *Example:* A loan officer will be responsible for processing a loan or credit application.

FA-10 Identify all operations performed on object types and relationships by the business activity.

> *Example:* Claims processing would create an occurrence of claimant and establish some type of relationship between claimant and policy holder.

END OF PROCEDURE: FUNCTIONAL ANALYSIS

4.3 ■ Scenario analysis

Scenario analysis uses a deep-background search for the relevant object types and relationships. It treats the domain of information as the setting for an active system. The search for components is based on the identification of the central activity of the system being modeled. Once that activity has been named, all participants are identified as possible object types (see Fig. 4.2).

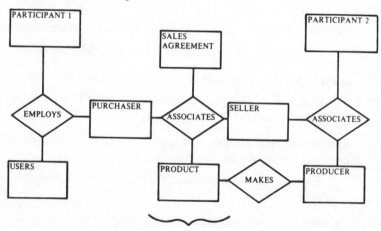

Central Activity of Domain

Figure 4.2. Scenario analysis.

Interactions between participants are searched for relevant relationships and mediating object types. Participants are analyzed for their organizational structure.

PROCEDURE: SCENARIO ANALYSIS

SA-1   Identify domain of information.

SA-2   Name and identify central activity of domain of information.

> *Example:* In Fig. 4.2, the hypothetical central activity would be selling.

SA-3   Identify all participating objects in the central activity.

> *Example:* Figure 4.2 indicates two participating organizations: a user organization, and a manufacturing organization. The selling activity is between these two participants.

SA-4   For each participating object type, identify all object types that describe the participant in some way, and identify the relationships between the descriptive entities and the participants. The descriptive object types would be concerned with describing time, place, or essential characteristics.

> *Example:* In Fig. 4.2, Participant 1, the user organization, may be physically distributed throughout the country. The respective physical sites may be described by some object type that specifies the physical location of each user group.

SA-5   Identify all relationships between participants, or between components of participants, that arise from the activity between the participants.

> *Example:* In Fig. 4.2, the object type SALES AGREEMENT associates PURCHASER, SELLER, and PRODUCT. The selling activity associates all these object types.

SA-6   Identify all object types that determine the nature of the interaction between participants.

> *Example:* In Fig. 4.2, SALES AGREEMENT defines the selling of PRODUCT to PURCHASER by the SELLER. Here is a case in which an object type defines an interaction *and* the resulting relationship. Because of this, we call SALES AGREEMENT an *associative object type*.

SA-7   Decompose all participants into an organizational structure of component object types and identify structural relationships between components.

> *Example:* Participant 1 of Fig. 4.2 is decomposed into USERS and PURCHASER. Participant 2 is decomposed into SELLER and PRODUCER.

SA-8   Identify any organizational structures that include the participants and their component objects.

SA-9   Identify all relationships that define interactions or logical associations between the components of the participants.

SA-10 Identify all object types that are used to determine or describe relationships between two or more object types.

> *Example:* Participant 2 is an organization whose internal structure associates SELLER with PRODUCER, and is another example of an associative object type.

END OF PROCEDURE: SCENARIO ANALYSIS

4.4 ■ Transaction analysis

Transaction analysis is based on a state-transition model of a system being modeled (see Fig. 4.3). The change of system state 1 to state 2 is effected by some event, which appears in the model as one or more transactions. The term "transaction" is equivalent to the term "operation."

Operations have been defined as actions that change the state of the system or system model. But informally, operations can be described as actions that create new relationships between components of the system being modeled. Augmented entity-relationship diagrams show operations (ellipses in the diagrams) establishing relationships between object types.

Transaction analysis, therefore, studies the nature of the transaction or operation that changes the state of the system.

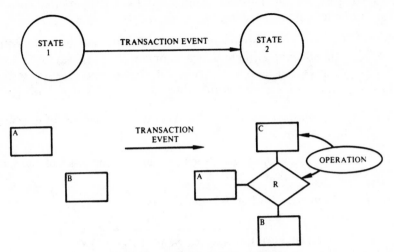

Figure 4.3. Transaction analysis.

PROCEDURE: TRANSACTION ANALYSIS

TA-1   Identify the domain of information.

TA-2   Identify all events that take place in this domain throughout time.

FOR EACH IDENTIFIED EVENT:

TA-3   Identify the transaction or transactions that caused the event to happen.

> *Example:* In Fig. 1.1 on page 8, the signing of a contract establishes an association between the SERVICE COMPANY, CLIENT COMPANY, and SERVICE PROJECTS.

TA-4   Name the user-defined operation that is implemented by the transaction.

> *Example:* As shown in Fig. 1.1, the named operation is SIGN CONTRACT.

TA-5 Identify all object types that are affected by the user-defined operation.

> *Example:* In Fig. 1.1, the object types are CONTRACT, CLIENT COMPANY, and SERVICE PROJECTS.

TA-6 Identify all relationships established by the operation between the affected object types.

> *Example:* Relationships may be directly defined in an information model, or indirectly implemented by way of an associative object type. Figure 1.1 shows how SIGN CONTRACT establishes a relationship through the object type CONTRACT.

TA-7 Identify all data elements whose content is changed by the user-defined operation.

TA-8 Identify all standard operations that are components of the user-defined operation.

> *Example:* See Fig. 1.1 and Exhibit 3.

END OF PROCEDURE: TRANSACTION ANALYSIS

These three search procedures are used for obtaining a first cut at an information model, and have great heuristic value in starting the search process. However, the model produced by these procedures is not in a refined state.

In order to refine the components of the model, procedures and rules are developed in Chapters 5 through 8. Chapter 5 develops the definition procedures for object types and the method of determining the data content of object types. Chapter 6 is important because it is concerned with detailed procedures for splitting out refined object types. This procedure is called *abstraction analysis,* and facilitates refinement of the set of object types discovered by the search procedures discussed in this chapter. Chapter 7 provides rules for refining the information model as a whole. And Chapter 8 refines the intuitively defined relationships of the first-cut model into one or more well-defined relationships. This procedure is called *anchor-point analysis.*

5

Declaration
of object
types

The three analytical procedures discussed in Chapter 4 are used to name and identify possible object types as components of an information model; but the naming of an object type is not sufficient justification for including it as a legitimate model component. This chapter is concerned with the problem of deciding when a candidate object type can be justified as a legitimate component of some information model. There are three basic criteria for declaring an object type as a component of some information model:

- relevance of the candidate object type to the domain of information

- definability of the function of the object type within the domain of information

- sufficient information content of the object type within the domain of information

All three criteria must be satisfied in order for an object type to be formally declarable within an information model. The term "formally declarable" means that the object type is described in the analytical partition of the information model. Exhibit 1 in Chapter 1 is an example of a formal declaration of an object type.

It is possible for an object type to be *referenced* in an information model without being formally declared. For example, one of the data elements attributed to the object type CONTRACT in Exhibit 1 is called SALESPERSON NAME. This data element could refer to the object type SALESPERSON without the object type appearing as a formal entry in the analytic partition of the information model. The hypothetical reason for this would be that, in Exhibit 1, SALESPERSON failed to satisfy the three criteria listed above.

The database partition of an information model stores information about object types in the form of data elements. If no data elements are attributed to an object type (that is, no information is maintained for that object type), then the object type is not formally declared in the model.

5.1 ■ Definability of object types

The most important aspect of a well-defined object type is that it can be properly defined *within* the system being modeled. Creating object type definitions is the first task of the model designer. The definition can be stated in terms of the object's characteristics. The first design decision rule appears below, and states precise requirements for proper definition of an object type's function.

RULE 1: *well-defined function*

Each well-defined object type *must* correspond to a distinct, definable *function* of some component of the system being modeled. The function is *definable* if it is possible to state its purpose, defining properties, *and* some combination of the remaining object characteristics. Its function must remain valid throughout the lifetime of the object type within the system being modeled.

Exhibit 6, on the next page, is an example of a well-defined function for an object type called PROJECT, which is implemented in a contractual context.

EXHIBIT 6

COMPONENT: OBJECT TYPE
<u>NAME</u>: SERVICE PROJECT
<u>DEFINITION</u>:

PURPOSE: > *To complete and deliver a unit of work for*
a contractual obligation.

PROPERTIES: > *A service project has one manager at any time.*
> *It employs members of the staff of the service company.*
> *A service project may also employ subcontractor personnel.*
> *It is conducted at one physical location.*
> *It has a termination date.*
> *It has a fixed set of deliverables.*

SYSTEM
EFFECTS: > *A service project is used to calculate charges.*
> *It can be delayed by other service projects.*
> *It can be prematurely terminated.*

OBJECT
EFFECTS: > *A service project may impact other contractual obligations.*
> *It can impede other service projects.*
> *It uses human resources.*

ASSOCIATIONS: > *A service project is subordinated to a specific contract.*
> *It receives deliverables from other projects.*
> *It forwards deliverables to other projects.*

The approach described in Exhibit 6 has made *function* the distinguishing characteristic of object types. Smith and Smith have argued that choosing a unique name for an object type makes it a legitimate entry in the information model. However, choosing names for object types does not guarantee that the model will accurately reflect actual policy as it governs the system being modeled. It is possible to select names for object types in the information model that make sense to an arbitrary observer, but would not represent anything recognizable in the domain of information. Also, it may be possible (but perhaps unlikely) to define a real function in the modeled system for which there is no appropriate English name. For these reasons, we shall use the object's function to define its type. For any object type that has a relevant well-defined function, *it is also necessary that it have a unique name, although it need not be an English noun.*

5.2 ■ Attributing data elements to object types

An object type is a component of the system being modeled about which facts are retained. These facts are called *attributes*.

DEFINITION: **attribute**

an attribute of an object type is a named characteristic of the object type such that for any single occurrence of the object type the attribute associates one or more data values with that occurrence. For example, if the object type is EMPLOYEE, then BIRTHDATE would be a single-valued attribute, and CHILD NAME could be a multi-valued attribute.

A problem that appears to be tangential to the problem of defining distinct object types has to do with the assignment of attributes. From a policy analysis point of view, the set of relevant object types in any system does not depend on their descriptive attributes. However, from a practical point of view, there is little cause to represent an object in an information model if we do not intend to keep any information about that object in the model. The information content of an object type in an information model can be identified with the set of its attributes. Therefore, the designer has two questions to resolve: Which objects are relevant to the system being modeled? Which subset of those objects should be represented in the information model?

The problem of sorting distinct object types, which we have begun to investigate, has substantial depth to it. The subsidiary problem of assigning attributes could easily be the subject of a separate treatise.

Third normal form* has been an attempt to provide a reliable method for resolving this problem, but it has its limitations.

Here is a rule (albeit non-mathematical) that can serve as a guideline for deciding whether an attribute fits an object type. The guideline for assigning attributes is: *An attribute is associated with a given object type only if it can serve as a partial specification of the function of the object type, for all occurrences of that object type.*

In the banking example, DAILY BALANCE can surely serve as a partial specification of the function of all DEPOSITORS. But it cannot serve to specify the function of all BANK CUSTOMERS, since some of them may not have accounts to which this attribute would apply.

An important special case of the attribute assignment problem concerns the unique identity of distinct occurrences of an object type. Is there a way to distinguish unique occurrences? The following rule addresses this case.

RULE 2: *unit identification*

For each object type, there must exist an attribute, or combination of attributes, whose values uniquely identify distinct occurrences of the object type. Rule 2 guarantees the existence of a *primary key* for each object type.

Another special case of the attribute assignment problem concerns attributes of object types and their subtypes. Let a *generic hierarchy* be the set containing some object type, all of its subtypes, all of the subtypes of its subtypes, and so on. For example, Fig. 3.6 shows a generic hierarchy. Consider the attribute STREET ADDRESS. It could be attributed either to BANK CUSTOMER or to any of its subtypes. Depending on where we start, this data element could propagate up or down the hierarchy. Let us adopt the convention that it be attributed to the highest object type in the hierarchy in which it occurs, and to that object type only. Summarizing, we state the third decision rule, known also as the attribution rule.

---

*Third normal form means that the attributes of an object type are functionally dependent on the unique identifier of the object type, the whole unique identifier, and nothing but the unique identifier of the object type.

RULE 3: *functional attribution*

A data element is attributed to a specific object type only if it can serve as a partial specification of the function of the object type, for all occurrences of the object type. Rule 3 is further strengthened if the following conditions hold:

☐ For each occurrence of the object type, the value (or values) of its attributes is exclusively determined by the full value of the primary key.

☐ For the case of the generic hierarchy in which a data element may be attributed to any object type in the hierarchy, the data element will be attributed to the highest object type to which it applies, and to that object type only.

The second condition of Rule 3 eliminates needless redundancy. It does *not* imply that the attributes of the highest object type in a generic hierarchy cannot logically describe lower object types in the hierarchy.

At this point, I shall invoke a regulation first enunciated by Codd of IBM. This regulation is called the Property Inheritance Regulation, and states that given any subtype e, all of the properties of its parent type(s) are applicable to e.

The attribution rule is a very powerful device for assuring that object types are described by their true characteristics, and *only* their true characteristics. It greatly enhances the modularity of the information model. The attribution rule reduces redundancy in an informal way that is analogous to third normal form, which reduces redundancy in a formal way.

5.3 ■ Formal declarability within an information model

An object type is represented by a data structure, in the sense that it has an object name and a primary key, and it bundles together a set of characteristic attributes. Certain identified object types in a given domain of information do not require a data structure for their representation. For example, for the EMPLOYEE object type in a personnel database, DEPARTMENT may be an attribute of EMPLOYEE. Suppose that there were no further characteristics of DEPARTMENT in the domain of information (that is, no more information about DEPARTMENT is of interest). The object type DEPARTMENT is represented by a single data element that is attributed to EMPLOYEE. The data element is an *identifier* for the DEPARTMENT object type. Let us define our terms:

DEFINITION: **identifier attribute**

an attribute X of an object type Y is an *identifier of Y* if there is a unique value of X for each distinct occurrence of Y, *and* there is public policy that recognizes X as a key of Y. Any attribute of Y that does not fit this description is called a *non-identifier attribute of Y*.

Object types can be represented in an information model in two ways: by their identifiers alone; or by their identifiers *and* non-identifier attributes (that is, their data structures).

RULE 4: *formal declarability*

An object type is *formally declarable* as an object component of an information model for a given domain of information if both of the following conditions are true:

☐ Rules 1 and 2 are obeyed.

☐ After Rule 3 has been applied to this object type, there is at least one non-identifier attribute associated with the object type.

Rules 1 through 4 are used to identify formally declarable object types that are well defined. Using these rules, the Property Inheritance Regulation, and the definitions in this chapter, we are able to distinguish formally declarable object types from among the potential object types for an information model. Our next concern will be determining at which level of detail a given object type should be declared.

Summarizing, we can say that an object type is formally declarable in an information model if it has a well-defined function in the system being modeled, and it is represented in the database partition by more than its unique identifier alone. In this chapter, we have required that each object type be well defined in terms of policy documentation. In Chapter 8, we shall require the same criteria for relationships.

6

Abstraction of object types

In this chapter, we address a very practical problem in database design: At what level of detail are object types properly declared? The problem is one of choosing between generality and detail, and is often a major obstacle for the beginning database designer.

Often the problem is caused more by prejudicial attitudes than by lack of technical judgment. Beginning database designers often make the mistake of deciding that an information model must have only one level of detail. Experienced designers know that no fundamental law is broken if multiple levels of abstraction are present in the same information model. Using the example of the BANK CUSTOMER, the subtypes BORROWER and DEPOSITOR can co-exist in the same model as BANK CUSTOMER without any fundamental problems as long as all object types are formally declarable.

The designer of an information model must learn how to spot situations in which abstracted object types are needed. In general, these situations are signaled by a lack of fit between declared object types and known policy, or by information gaps in the resulting model.

6.1 ▪ Types of abstraction

In Chapter 3, four modes of abstraction were cited. For each mode there is a corresponding decision rule to guide the analyst/designer in applying the abstraction principles.

Table 6.1
Abstraction Mode/Decision Rule Correspondence

Abstraction Mode	Decision Rule
functional differentiation	Rule 5: *subtype independence*
classification	Rule 6: *supertype independence*
association	Rule 7: *aggregate independence*
characterization	Rule 8: *context independence*

The following three subsections elaborate on the correspondence between Rules 5 through 8 and the abstraction principles.

6.1.1 ○ Classification and differentiation

One of the most common problems experienced by logical database designers is the failure to differentiate between related, but functionally distinguishable, object types. An example is failing to distinguish between BORROWERS and DEPOSITORS in a banking database. The result is a loss of structural modularity and clarity, the production of update anomalies, and the fragmentation of update processing.

One indication that object types are not sufficiently differentiated and that fragmentation exists occurs when, for any object type X, some update transactions operate on a distinguishable subset of the occurrences of X, while other update transactions operate on some other disjoint subset. For any declared object type, the analyst must always ask whether the object type plays two or more distinguishable roles in the system being modeled. Before giving the rule for functional differentiation, I shall provide two pertinent definitions.

DEFINITION: **object subtype**

for any given object type, an *object subtype* is a distinguished category, or subset, of all individuals represented by the object type. It is generally derived by *functional differentiation* of the original object type.

DEFINITION: **object supertype**

for any object type, its *object supertype* is any object type that includes it as a subtype. It is generally derived by a classification of the original object type.

The entity-relationship diagram in Fig. 6.1 shows a generic hierarchy of supertypes and subtypes. Supertypes and subtypes form a generic family around any given object type.

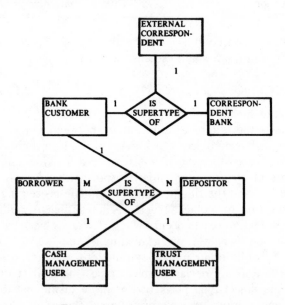

Figure 6.1. A generic hierarchy.

At what stage do we identify independent subtypes? In the case of our BANK CUSTOMER example, the subtypes BORROWER and DEPOSITOR are rather clearly distinguished. But do we need to formally declare them?

In the example of a personnel management database, there is an object type called ASSIGNMENTS. Suppose that for purely political reasons some ASSIGNMENTS are called PROJECTS, others are called TASKS, and still others are just called ASSIGNMENTS. Under which conditions do we formally declare TASKS and PROJECTS as independent object types? The following rule helps us address this problem:

RULE 5: *subtype independence*

For any object type A, we formally declare as independent subtypes A1 and A2 if the first two conditions listed below are both true, *or* if the third condition is true:

☐   Rules 1 through 4 are obeyed by subtypes A1 and A2.

☐   After application of Rule 3, there is at least one non-identifier attribute associated with one of the subtypes, but not the other.

☐   There exists explicit public policy that defines two disjoint subsets of A as A1 and A2, respectively, and assigns each subtype unique values and disjoint sets of primary keys.

A less common but frequent problem occurs when a database designer fails to take advantage of the concept of *generality*. Suppose that our banking database counted the following object types: BORROWER, DEPOSITOR, CASH MANAGEMENT USER, and TRUST MANAGEMENT USER. Suppose also that there were common data that described each of these object types equally well. To which of the object types would these data be attributed? Should the common data be attributed to each of these object types in a redundant fashion?

The correct action is to declare BANK CUSTOMER as a supertype of the other object types. This leads us to Rule 6:

RULE 6: *supertype independence*

For any set A of objects X1 . . . Xn, we formally declare A to be an object type if both of the following conditions are true:

☐   Rules 1 through 4 are obeyed by A.

☐   After application of Rule 3, there is at least one non-identifier attribute associated with A, but not with the objects X1 . . . Xn.

6.1.2 ○ Association

Object types are usually associated by relationships, but often an association is defined by an object type in the system being modeled. What kinds of object types associate other object types? Some examples are listed below:

- social organizations such as clubs, teams, or corporate employers

- social agreements such as legal contracts, oaths, or social covenants

- transaction documentation such as purchase orders, waybills, or bills of lading

Figure 6.2 provides an example of a document, ENROLLMENT, that associates STUDENT with CLASS. This relationship is defined below.

DEFINITION: **associative object type**

an object type is called *associative* if it defines a specific relationship between two or more distinct object types. An associative object type is often called an *aggregate*.

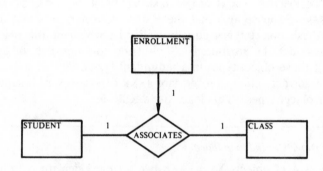

Figure 6.2. Associative object types.

Now that we have defined and illustrated associative object types, we need to refine our terms by further specifying the associative relationship. A definition of *constituent object type* will do this.

DEFINITION: **constituent object type**

any object type related to other object types by way of some associative object type is a *constituent* of that associative object type. For example, in Fig. 6.2, STUDENT and CLASS are constituents of ENROLLMENT. Relative to the object type STUDENT, we say that ENROLLMENT is an *associative* object type.

The associative object type raises another design dilemma for the model designer: How does one distinguish between an associative object type and a relationship? Both formal structures associate two or more object types for some specific reason.

In order to settle this question, we must remember that an association between two or more object types is an abstraction from a specific kind of interaction between those objects in the system being modeled. If we say that a supplier *supplies* parts *to* a customer, we are describing a certain type of interaction (supplying parts) that occurs between SUPPLIER and CUSTOMER and involves PARTS. In relational notation, we could write: SUPPLIES (SUPPLIER, PARTS, CUSTOMER).

This relational notation emphasizes the point that the relationship is an abstraction from interactions between entities. Does the abstraction itself interact with other object types in the model? That is, can the associative abstraction enter into relationships (higher order) with other objects? If the answer is yes, then we can formally declare the associative abstraction as a new object type in the model. This is entirely consistent with the principle of object normalization.

For example, DEPARTMENT is an associative abstraction from the organizational interaction between MANAGER and EMPLOYEE. But DEPARTMENT can interact with other functional units such as COMMITTEE. These interorganizational relationships between functional units endow DEPARTMENT with the status of organizational entity within the COMPANY; that is, as an object type within the COMPANY information model. This is the essence of our next design decision rule.

RULE 7: *aggregate independence*

An associative abstraction from the interaction of several object types in the system being modeled is formally declared as an object type if both of the following conditions are true:

- ☐   Rules 1 through 4 are obeyed by this object type.
- ☐   This associative abstraction can itself interact with other object types in the system being modeled; that is, it is capable of entering into formal relationships with other object types.

6.1.3 ○ Characterization

The last form of abstraction to be dealt with is surely the most subtle. It treats a characteristic of some object type as an independent object type in its own right. When are we justified in doing this, and what are examples of characteristic object types? Let us turn first to our definition.

DEFINITION: **characteristic object type**

an object type X characterizes an object type Y if X describes Y in some way, if specific occurrences of X are attributed to specific occurrences of Y, *and if the non-existence of Y implies the non-existence of X.* An object type is *independent* if it is not a characteristic object type.

The most general justification for declaring characteristic object types is that they fulfill some function necessary to the domain of information. I can reformulate this justification by invoking a variation of the object normalization principle: *A characteristic of some object type can be abstracted into a characteristic object type if the characteristic itself possesses characteristics in the system being modeled.*

Some examples of characteristic object types are formal measurements of object types; time-dependent descriptive information; historical information; performance analyses; and projections, correlations, and simulations.

EMPLOYEE

SSN	NAME	ADDRESS	CITY
091 36 3694	J. Jones	10 Browing St.	London

CHILD

SSN	FIRST NAME	BIRTHDATE	SEX
345 26 2645	Peggy	9 Aug 74	F
287 49 4987	Derek	5 June 67	M

Figure 6.3. An object type with one of its characteristic object types.

Figure 6.3 shows how CHILD is a characteristic object type associated with EMPLOYEE. In this example, CHILD information certainly describes EMPLOYEE, but it does not appear particularly abstract. However, if EMPLOYEE were also described by a characteristic object type called DEGREES (type, date received, conferring school, and so forth), it would strike most readers as a more abstract object type than CHILD.

Most beginning database designers are perplexed by characterization. Characteristic object types always appear to the viewer to be much more abstract than the object types they describe. This leads analyst/designers to doubt their legitimacy within information models. Often the attributes of characteristic object types are attributed to object types that they do not truly describe, simply because the database designer is afraid to declare the characteristic object type as a legitimate component of the information model.

Characteristic object types are usually represented as *dependent segments* in the data structure of some formally declared object type. (Exhibit 1 in Chapter 1 serves as an example.) A dependent segment is the same as a repeating group in a standard record format. For example, the CHILD data in Fig. 6.3 would be represented as a dependent segment in the data structure of the EMPLOYEE object type. This conforms with the subordinate nature of characteristic object types.

However, some apparently characteristic object types should not be subordinated to other object types, but should be declared independently of any other object types in the model. The next decision rule can be used to guide the analyst/designer in deciding whether an object type is characteristic of some other object type or independent of other object types altogether. A definition of value-dependency follows:

DEFINITION: **value-dependency**

if A and B are two attributes of object type X, and A is single-valued for each distinct occurrence of X, then B is *value-dependent* on A if for each distinct value of A there is a finite, unique set of B values that corresponds to it.

From this definition and the definition of a characteristic object type, it follows that the attributes of any characteristic object type are value-dependent on the primary key of the owning object type. Value-dependency is the distinguishing mark of characteristic object types.

A subtle question for the designer is whether an object type should be subordinated to another object type as a characteristic if it describes the other object in some way. This is addressed in the following decision rule.

RULE 8: *context independence*

An object type is formally declared in the information model *independently of the context of any other object type* if the first condition given below is true and either or both of the second and third conditions are true:

☐ The object type obeys Rules 1 through 4.

☐ The attributes of the given object are *not* value-dependent on the primary key of any other object type in the model.

☐ The given object does *not* require the information context of any other object type in the model in order to be relevant to the domain of information.

6.2 ▪ Abstraction analysis

Following is an analytical procedure that summarizes the rules and techniques of abstracting object types. It is to be used for each candidate object type.

PROCEDURE: ABSTRACTION ANALYSIS

AA-1 Check that the object type has a well-defined function in the system being modeled. If so, apply Rule 1.

AA-2 Check that the object type is formally declarable within the information model. Apply Rules 1 through 4.

AA-3 Does the object type perform multiple subfunctions in the system being modeled so that each subfunction defines a subset of the object type that is disjoint from other subsets so defined? If so, declare the subtypes as independent object types in the model. Apply Rule 5.

AA-4 Does the object type share common attributes with other object types in the domain of information? Can all object types related in this way be classified as instances of a more general object type. If so, declare the supertype as an independent object type in the model. Apply Rule 6.

AA-5 Does the object type participate in some type of agreement, arrangement, or legal covenant? Apply Rule 7.

AA-6 Does the object type participate in some organizational structure, or social organization? Apply Rule 7.

AA-7 Is the object type associated with other object types by some transaction documentation? Apply Rule 7.

AA-8 Does there exist some type of descriptive profile, performance measurement, or simulation of the object type? If so, declare a characteristic object type. Apply Rule 8.

AA-9 Does there exist time-dependent or historical descriptions of the object type? If so, declare a characteristic object type, and apply Rule 8.

AA-10 For any given object type, is there a characteristic of the object type that itself possesses characteristics within the domain of information? If so, declare the characteristic to be a characteristic object type of the given object type. Again, apply Rule 8.

AA-11 For any abstracted object type, perform AA-1 and AA-2.

END OF PROCEDURE: ABSTRACTION ANALYSIS

Keep in mind that instances of abstracted object types pose serious implications for the model representation phase. They concern existence dependencies between various object types. For example, in some domains of information, the elimination of an object type implies the elimination of all of its subtypes. And, we have seen that when an object type is deleted or eliminated, all of its characteristic object types must be eliminated. In the case of associative object types, elimination of any constituent implies that the associative object type in the model could be in error.

Therefore, the richness of a model in the number of related abstract object types is offset by the complexity of the integrity constraints that are built into the model in the representation phase. The next chapter treats refinement of the object set, a process that minimizes the model's complexity.

The purpose of abstraction analysis is to refine the set of object types that were originally identified by search procedures. This refinement is concerned with the splitting out of more finely resolved types of objects. Within the context of information modeling, abstraction analysis can be considered a detailed analytical procedure.

7

Refinement
of the
object set

Up to this point, we have examined how to decide what should be formally declared as an entity or object in an information model. The terms *entity* and *object* have been left as intuitive notions, and in their place I have proposed the more precise notion of the functionally determined *object type*. The design decision rules have been oriented toward the identification of object types one at a time.

In this chapter, we shall consider the set of all object types declared in the information model. Two questions are pertinent: How do the object types work together to provide a conceptual model of a given domain of information? What global decision rules apply to the entire information model?

We shall address these questions by focusing on the following four design problems: how to determine the set of necessary object types, how to determine the sufficiency of the set of object types, how to derive a model with minimum redundancy and complexity, and how to ensure the modularity of the derived model.

7.1 ■ How to determine the set of necessary object types

Determining the specified set is our first problem and is directly addressed by the following decision rule.

RULE 9: *domain necessity*

An object type is a necessary component of an information model for a given domain of information if the first condition listed below is true *and* if either or both of the second and third conditions are true:

☐ The object type obeys Rules 1 through 4.

☐ It is impossible to accurately specify the contents of the domain of information without explicit reference to the object type.

☐ There are data elements in the data dictionary for the domain of information for which the attribution rule (Rule 3) associates them with the given object type, and no other.

The value of Rule 9 is that it can be used to eliminate object types ꞏidentified in the search process that are not essential to the domain of information.

7.2 ■ How to determine the sufficiency of the set of object types

This second problem deals with the sufficiency of the set of object types declared in the model. Before defining the tenth decision rule, it is necessary to consider the problem of intersection data.

DEFINITION: **intersection data**

one or more data elements taken from the data dictionary that are partially attributable to two or more object types in the model. That is, the values of these data elements are value-dependent on the *combination of* primary keys of two or more object types in the model.

Take as an example the data element GRADE, which is an attribute of the combination of STUDENT and CLASS. If there exists an object type ENROLLMENT in the domain of information, then GRADE would be a legitimate attribute of this associative object type. However, the educational environment may not have defined ENROLLMENT as a formal aspect of policy. Yet, the database designer cannot create fictions in the information model. The model must be a *faithful* rendering of the domain of information. Therefore, it seems in this case that we have no place to put GRADE in our model.

Whenever intersection data exist *and* there is no formally declarable associative object type to assume them, then it is necessary to introduce a special data structure to assume the intersection data. A definition of *intersection* follows.

DEFINITION: **intersection**

a data structure composed of a concatenated primary key and intersection data. The primary key is a concatenation of the primary keys of the object types on which the attributes depend. Each attribute is value-dependent on the full combination of the primary keys.

The data structures of an information model will include object types, relationships, attributes, and possibly one or more intersections. Intersections are included *only* if there are no associative object types to assume the intersection data.

Now we may turn to the rule for establishing the sufficiency of the set of declared object types.

RULE 10: *domain sufficiency*

The set of formally declared object types comprising an information model for a given domain of information is sufficient if both of the following conditions hold:

☐ All domain-necessary objects are present in the information model.

☐ The object set corresponds in content to the material documented in the data dictionary; that is, each data element in the data dictionary is attributable to some object type in the model, *or* to some *combination* of object types in the model.

7.3 ■ How to derive a model with minimum redundancy and complexity

Based on the notions described above, we can prescribe a rule for deriving a model that has minimal complexity.

RULE 11: *minimal complexity*

The object set of an information model is judged to be of minimal complexity if the object set is domain-sufficient, *and* the *only* formally declared object types are domain-necessary.

The minimal complexity rule corresponds to an all-and-only condition holding on the formally declared object types of the information model. We can apply a similar all-and-only rule to the distribution of attributes.

RULE 12: *modularity*

The object set of an information model is modular if no object type contains attributes whose values are totally or partially value-dependent on the primary key of some other independent object types.

In effect, Rule 12 tells us that each object type is characterized by attributes that are totally value-dependent on the primary key of the object. We have now seen the first twelve rules. Let us briefly review them before turning to the next chapter and a presentation of the techniques of anchor-point analysis.

Rules 1 through 9 guide the designer on deciding if a candidate should be formally declared as an object type in the model. Rules 10 through 12 assist the designer in a final evaluation of the object set in order to create a model with structural economy. Furthermore, Rules 9 through 12 play an important role in the definition of the well-formed model, which is formally defined in Part Three.

8

Generalized relationships

Traditionally, analysts have restricted relationships in databases to the binary case; that is, as associations between two and only two object types. Information modeling takes a decisive step in generalizing the concept of *relationship* in order to define associations between any number of distinct object types, with binary relationships as special cases of n-ary relationships.

The reason for instituting generalized relationships is that the real world is not restricted to binary associations. For example, CUSTOMERS, SUPPLIERS, and PARTS are all associated in a natural way that is familiar to us. Here we have three object types. Of course, we could present this association in purely binary terms: CUSTOMER SUPPLIER, CUSTOMER PART, PART SUPPLIER. But these three binary relationships can be just as easily understood in terms of one tertiary relationship: CUSTOMER PART SUPPLIER. Therefore, generalized n-ary relationships represent real-world events and associations in a more natural way than do binary relationships.

However, this increase in naturalness and fidelity to the real-world system is balanced by a corresponding increase in the complexity of specifying an n-ary relationship. Generalized relationships are more complicated.

Binary relationships are frequently analyzed in terms of the usual litany of combinations: one-to-one, one-to-many, many-to-one, and many-to-many. Generalized relationships greatly increase the complexity of this combinatorial analysis. Indeed, this type of combinatorial analysis is not practical with n-ary relationships.

In the absence of combinatorial analysis, we are faced with the problem of how to define n-ary relationships in a precise way. The technique that solves the problem is called *anchor-point analysis,* and is developed in the following sections.

8.1 ■ Relationships as abstractions

A relationship is the *result* of some real-world process that brings together a set of participating object types. Because a relationship is an abstraction, it does not describe the process that brought it about. A relationship only describes the ways that the participating object types are combined, or associated, as a result of the existence of the real-world process. Let us now formally define *relationship.*

DEFINITION: **relationship**

an association defined between the occurrences of two or more object types is a relationship.

As we have stated before, a relationship is an abstraction from interactions between object types in the system being modeled. In fact, relationships are generally derived from the following situations:

- logical associations resulting from some form of abstraction between object types — for example, generic relationships between objects and their subtypes, or associative relationships between objects and their constituents

- interactions between object types in the system being modeled; these interactions are specified by the rules, policies, or laws that govern the activity of the system

Looking back at Chapter 3, we can see that Figs. 3.4 through 3.6 are examples of relationships that are based on logical associations between object types. Figure 8.1 below shows a relationship based on business interactions between CUSTOMERS and the SUPPLIERS who supply them with PARTS. The underlying interpretation of Fig. 8.1 is: "Each supplier supplies multiple customers, each with one or more distinct types of parts."

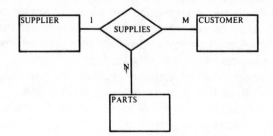

Figure 8.1. A relationship based on business interactions.

Noting how relationships are derived brings us to a major consideration concerning the definition of relationships, and is the subject of the next section.

8.2 ■ Definition of relationships

A relationship between two or more object types is definable if it is based on the *policies, rules, conditions,* or *laws* that govern the object types and their interactions in the system being modeled. For example, the relationship SUPPLIES in Fig. 8.1 expresses the business policy that a SUPPLIER is permitted to supply more than one part to each CUSTOMER.

Figure 8.2. The effect of a condition on a relationship.

Clearly, Fig. 8.2 shows a relationship LEGAL FATHER OF, which is *conditioned* in the sense that it is true only if the object type MAN is legally married to the mother of the child or children referenced by the object type CHILD.

Just as it is necessary to define the object function of each object type in the information model, it is necessary to specify the system-determined rules that define relationships between object types.

EXHIBIT 7

COMPONENT: RELATIONSHIP
NAME: LEGAL FATHER OF
DEFINITION:
    BASIS: > *For each man, there are child(ren) such that man is the legal father of the child(ren) if the referenced man is legally married to the mother of the referenced child(ren).*

We are now ready to state our next design decision rule, Rule 13.

RULE 13: *associative basis*

A relationship between N object types is formally declarable in an information model only if the following two conditions are both true:

☐ It is possible to define the rules of association between distinct occurrences of each of the N object types in terms of known laws or policies that govern the referenced object types within the system being modeled.

☐ It is possible to state any and all conditions that are necessary to validate an association determined by known laws and policies.

We can record the associative basis for the relationship depicted in Fig. 8.2 to produce Exhibit 7, shown on the preceding page.

8.3 ■ Well-definedness of relationships

This section develops a definition for these qualities of a relationship that constitute its being well defined. We shall refer to this quality as the relationship's "well-definedness." It is possible to show that well-definedness can be accomplished by means of anchor-point analysis, the analytical procedure fully presented in Section 8.4. If anchor-point analysis is used to specify an n-ary relationship, then that relationship will be well defined.

In order to establish this, it is necessary to develop the underlying theory, based on the Binary Equivalence Theorem. For those of you who are interested in the development of the theory, I urge you to read this section. Those of you whose interest is primarily the resulting analytical procedure may proceed to Section 8.4.

Despite the power of Rule 13 to make relationships precisely definable, there still remains one last problem to be resolved before a relationship can be defined with complete precision: Given the complexity of n-ary relationships, can we precisely determine which occurrences of each object type are associated with distinct occurrences of each of the remaining object types in the relationship? We can, but the solution is somewhat involved.

The ultimate solution is usually achieved by defining the n-ary relationship in terms of an exhaustive n × m table of unique identifiers, where n represents horizontal rows and m represents vertical columns. Table 8.1 defines the relationship SUPPLIES.

Table 8.1
Relational Table for SUPPLIES

SUPPLIES

SUPPLIER	PART	CUSTOMER
INTEL	MICROCOMPUTER	DUPONT
INTEL	RAM	BUNKER-RAMO
INTEL	ROM	MATSUSHITA ELECTRIC
FAIRCHILD	MICROPROCESSOR	GENERAL ELECTRIC
FAIRCHILD	RAM	UNITED TECHNOLOGIES
TEXAS INSTRUMENTS	ROM	DUPONT
TEXAS INSTRUMENTS	RAM	W.R. GRACE
TEXAS INSTRUMENTS	MICROCOMPUTER	EXXON

Each row in Table 8.1 defines an *occurrence* of the relationship SUPPLIES. Each occurrence associates specific occurrences of each of the participating object types. Three object types are present: SUPPLIER, PART, CUSTOMER. The columns contain unique identifiers (primary keys) for any object type participating in the relationship; the set of unique identifiers for all occurrences of the object type is called the *domain of unique identifiers*. The number of columns in a table corresponds to the number of domains of unique identifiers. The number of domains need not be distinct. A formal definition follows.

___

DEFINITION: **relational table**

a table consisting of unique identifiers for a set of associated object types. There is one column for each object type participating in the relationship. An object type may occur in more than one column of the table. All entries in a given column are drawn from the domain of unique identifiers of the object type represented by that column. Each row defines an occurrence of the relationship, which associates distinct occurrences of each of the participating object types. A relational table is *complete* if all entries in the table have non-null values.

___

The main value of a complete relational table is that it provides a total specification of a relationship. However, its main drawback is that it is often very large. In real-world databases, there may be several thousand or several million occurrences of the relationship (rows in the table). This is allowable for storage purposes, but is really very inefficient for *specification*. To use a relational table for specification of a relationship is to perform the task the hard way.

Is there an easier way to specify a relationship with precision? Also, can we define what we mean by a precise specification? The answer to both questions is yes.

DEFINITION (PRELIMINARY*): **well-defined relationship**

a relationship between object types is well defined if for any occurrence of any participating object type it is possible to associate it *unambiguously* with specific occurrences of any other participating object type.

It is easy to demonstrate that if there is a complete relational table that specifies a relationship R, then R is well defined. But there is an easier way to guarantee that R is well defined — and that way is through the methods of anchor-point analysis.

8.3.1 ○ Specifying associations

Anchor-point analysis states that we specify associations between each of the participating objects in some n-ary relationship R, in terms of a fixed referenced object type called the anchor object. Let's use an example. Figure 8.3 shows a four-way relationship between CONTRACT, SALESPERSON, CONTRACT OFFICER, and PROJECT. We choose CONTRACT as our anchor object because it will determine the occurrences of the other object types that are associated with it. Note that our anchor object is the subject of the relational statement that captions Fig. 8.3.

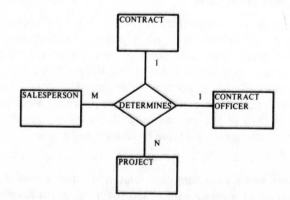

Figure 8.3. CONTRACT DETERMINES the responsible SALESPERSON, CONTRACT OFFICER, and subordinated PROJECTS.

---

*The qualifier PRELIMINARY is used on those definitions for which final definitions will be presented in Part Three of this book.

Once we have selected an anchor object, we then define the association between occurrences of each pair of participating objects. For the relationship DETERMINES in Fig. 8.3, there are six pairs of object types:

CONTRACT , SALESPERSON
CONTRACT , PROJECT
CONTRACT , CONTRACT OFFICER
SALESPERSON , PROJECT
SALESPERSON , CONTRACT OFFICER
PROJECT , CONTRACT OFFICER

If a relationship R has N participating object types, there will be $N(N - 1)/2$ distinct pairs of object types for R. We specify the association between occurrences of two participating object types if they have a common association with some occurrence of the anchor object.

DEFINITION: **anchored binary relationship**

a relationship between two object types such that the association between the occurrences of the object types is determined by a common association to occurrences of some anchor object. Each occurrence of either object type must be associated with some occurrence of the other object type.

Does anchor-point analysis yield well-defined relationships? The Binary Equivalence Theorem and its accompanying proof, given on the following page, show us why the answer is yes. Note that the theorem itself is presented first, followed by the formal proof, consisting of development of the IF portion and then the ONLY IF portion. Our proof brings us to Rule 14, the last rule for relationships.

RULE 14: *binary associativity*

Each relationship R between N object types, which is formally declared in an information model, must satisfy the following conditions:

☐ It must obey Rule 13.

☐ It must be defined by the specification of the $N(N - 1)/2$ anchored binary relationships that are associated with R.

BINARY EQUIVALENCE THEOREM

The theorem for binary equivalence can be stated as follows:

A relationship R between N object types is well defined if and only if *there exist N(N − 1)/2 anchored binary relationships, R_{ij}, that specify associations between each pair of object types referenced by R.*

PROOF: IF part of the theorem

In order for R to be well defined, it is necessary to show that the anchored binary relationships, R_{ij}, unambiguously can associate any occurrence, X, of any participating object type, O_k, with occurrences of any other participating object types. Choose some other object type, O_m. By hypothesis, there is an anchored binary relationship, R_{km}, that associates occurrences of O_k with occurrences of O_m. Each occurrence of O_k is associated with specific occurrences of O_m on the basis of a common association with one or more occurrences of the anchor object. It follows that the association of X with occurrences of O_m is unambiguous. Since O_m is arbitrary, it follows that R is well defined.

PROOF: ONLY IF part of the theorem

In order to show that $N(N − 1)/2$ anchored binary relationships exist, it is necessary to show that R can be decomposed into binary R_{ij} that are anchored, given that R is well defined. I shall make use of the fact that if relationship is well defined, then it is possible to construct a complete relational table that represents the relationship. Let such a table exist for R. It has n columns. Now, choose an anchor object for R in the following way: Each column of R contains values from the domain of unique identifiers for the object type associated with that column. If there are m rows in R, then the number of distinct unique identifiers in any column is less than or equal to m. Let K_i be the number of distinct unique identifiers in the i-th column of R. Then there are n numbers, $K_i \ldots K_n$. One of these numbers, K_s, will be less than or equal to all the others. Let the object type associated with column s be the anchor object. Form R_{ij} from the i-th column of R and the j-th column of R, with matching rows as in R. Occurrences of O_i are associated with occurrences of O_j by virtue of being in the same row with occurrences of O_s. Each specific occurrence of O_i in R is associated with some occurrence of O_j in R, and vice versa. Therefore, R_{ij} is anchored, and the theorem is proved.

8.4 ■ Anchor-point analysis

According to the Binary Equivalence Theorem, a necessary and sufficient condition for proving a relationship to be well defined is that all pairs of participating object types be associated by *anchored binary relationships*. A pair of object types is associated by an anchored binary relationship if each occurrence of the first object type is associated with some occurrence of the second object type by virtue of the fact that both occurrences are associated with a single, common occurrence of a third object type, the anchor object. This means that A and B, for example, are associated together *if* both A and B are associated with C.

If there are N object types participating in a relationship, there will be $N(N - 1)/2$ pairs of object types. However, *we can reduce the number of binary specifications to N, if N object types participate in a relationship.* Returning to Fig. 8.3, the relationship DETERMINES associates the object types CONTRACT, SALESPERSON, CONTRACT OFFICER, and PROJECT. Since there are four object types, then $4(4 - 1)/2 = 6$. The resulting six pairs of object types are

CONTRACT , SALESPERSON	SALESPERSON , PROJECT
CONTRACT , PROJECT	SALESPERSON , CONTRACT OFFICER
CONTRACT , CONTRACT OFFICER	PROJECT , CONTRACT OFFICER

Note that if we choose CONTRACT as our anchor object, then there are three relationships that associate CONTRACT with the other object types. These are the three in the left-hand column of the list shown above. It is always true that of the $N(N - 1)/2$ possible pairs of participating object types, $N - 1$ of those pairs will associate the anchor object with one of the remaining participating object types.

By definition, the first three relationships shown above are anchored binary relationships. Having established these $N - 1$ explicit anchored binary relationships, we must specify that the remaining relationships are anchored on CONTRACT. Note that these remaining binary relationships do not contain the anchor object. The rule for associating two non-anchor object types is simple and is derived from our definition of anchored binary relationship:

Associate A and B if *both* A and B are associated with C (the anchor object).

This single rule will cover all remaining cases. So, in order to specify $N(N - 1)/2$ anchored binary relationships, it is necessary to define $N - 1$ binary relationships between the anchor object and the remaining $N - 1$ object types, *and* to specify the general rule for associating the occurrences of pairs of non-anchor object types.

For a relationship with N participating object types, only N binary specifications are needed to ensure that all $N(N - 1)/2$ pairs of participating object types are associated by anchored binary relationships.

EXHIBIT 8

COMPONENT: RELATIONSHIP
NAME: DETERMINES
DEFINITION:

BASIS:
> For each contract, there is a contract officer,
and there are salespersons and projects such
that these objects are associated if the salespersons
participated in the sale of the contract
and the contract was signed by the contract
officer for the rendering of service projects.

BINARY
ASSOCIATIONS:
> For each contract, there are salespersons such that
the salespersons participated in selling the contract.
> For each contract, there are projects such that
the contract defines service projects to be
rendered to a client company.
> For each contract, there is a contract officer
such that the contract officer participated in
closing the contract by signing the contract.
> For each contract, there is a contract officer,
and there are projects and salespersons such
that objects A and B are associated if A, B
represent a pair from (contract officer, projects,
salespersons), and A, B are both associated
with the same contract.

Please note that we are *not* converting an n-ary relationship to N binary relationships. However, we are adding N specifications to our relationship definition in order to ensure that the relationship is well defined.

Exhibit 8 provides a complete specification of the relationship DETERMINES, which we saw in Fig. 8.3. Under the heading BINARY ASSOCIATIONS, there are N = 4 entries. The last entry specifies the general rule for anchored binary relationships.

According to the Binary Equivalence Theorem, the specification given in Exhibit 8 is a well-defined relationship.

8.4.1 ○ Anchor-point analysis procedure

Anchor-point analysis is a procedure for identifying and defining well-defined generalized relationships. The steps for this procedure are given below.

PROCEDURE: ANCHOR-POINT ANALYSIS

AP-1   Identify any generic relationships occurring between subtypes and a supertype.

> *Example:* There is a generic relationship between BANK CUSTOMER and BORROWER and DEPOSITOR.

AP-2   Identify any characteristic relationships between one independent object type and its characteristic object types.

> *Example:* There is a characteristic relationship between EMPLOYEE and SECURITY CLASSIFICATION if the latter is a formally declarable object type.

FOR ANY EVENT, TRANSACTION, OR OPERATION:

AP-3   Identify all distinct associations that result from interactions between object types brought about by events, transactions, or operations.

FOR ANY IDENTIFIED RELATIONSHIP:

AP-4   Select the object type, from all participating object types, on which the relationship can be logically based. This object type becomes the anchor object.

In general, this selection is based on the following criteria:

○ This object type helps to define the association between participants.

○ No definition of the relationship can be given without referring to this object type.

○ This object type is the most prominent component of the underlying policy.

○ This object type varies the least quickly in combination with the remaining participants.

AP-5 Define the associative basis for the relationship by stating the rules, laws, and conditions that determine the association of the participating object types. State this definition in terms of the anchor object.

Example: See Exhibit 8.

AP-6 Define the N − 1 binary relationships that exist between the anchor object and the remaining N − 1 non-anchor object types. Maintain consistency between the associative basis and these definitions.

Example: See Exhibit 8.

AP-7 Define the general rule for associating all pairs of non-anchor object types in anchored binary relationships.

Example: Follow the last entry in Exhibit 8.

END OF PROCEDURE: ANCHOR-POINT ANALYSIS

8.5 ■ Concluding comments

Anchor-point analysis is a refinement of relationships in the model in much the same way that abstraction analysis is a refinement of object types. Both procedures are used to refine the entity-relationship components of an information model. After these procedures have been performed, the resulting model should be verified with the user community.

Part Three
Conclusion

Part Three consists of a summary of the basic premises and fundamental semantic concepts of information modeling. In Chapter 9, the final definitions of well-defined object type, well-defined relationship, and well-formed model are given, based on the complete set of rules established in Part Two. In Chapter 10, core analytical procedures are summarized, and the analysis and representation phases of information modeling are set forth in a step-by-step fashion.

9

Summary of concepts and final definitions

Information modeling places great emphasis on the *techniques* of analysis. It tells the analyst how to analyze systems and build models. The tools of information modeling include semantic concepts that define what components the analyst should use in building models. Part Two of this volume built the logical foundations for the following semantic concepts:

- well-defined object types
- well-defined relationships
- well-formed models

This chapter provides a summary of the definitions and rules assembled in Part Two — the logical foundations — and the final rigorous definition of each design concept.

9.1 ■ Well-defined object types

The distinguishing feature of an entity is its *characteristic role* or *function* in the system being modeled. This characteristic function serves as an entity's *identity* in the system being modeled. Because an entity may possess more than one characteristic function or role in the system, the fundamental semantic component of an information model is the *object type* rather than "object" or "entity." Each distinct role or function of an object determines a distinct object type.

Introducing the object type clears up a fundamental problem of representation in an information model, raised in Chapter 1: How do we represent the fact that a single entity (BANK CUSTOMER) has four different roles in the system being modeled (BORROWER, DEPOSITOR, TRUST USER, CASH USER)? Each function becomes a separate object type, and in our example, BANK CUSTOMER also becomes a declarable object type. In this example, five object types have been declared: one supertype and four subtypes. The semantic notion of the *object type* resolves basic ambiguities in the structure of information models.

Another set of problems addressed by the notion of object type has to do with the level of detail represented in the model: differentiating between general versus detailed information. These problems revolve around the process of abstraction, the process of deriving new object types from old ones, which we discussed in Chapter 6. Often, this means that general information, such as supertypes and associative object types, is abstracted from detailed information. It is also possible to proceed from general to detailed, in the case of deriving subtypes from a proposed object type. The essence of abstraction is that the observer (user or analyst) *selects* a set of attributes or characteristics that describe an object type at a different level from that of the object types previously defined. That level can be more general or more detailed, but it is derived from prior object types.

In addition to the idea of observer selection, information modeling has added another process to the definition of abstraction. *Characterization* is closely related to the basic psychological processes of conceptualizing the real world; a common feature of ordinary language, it consists of treating the characteristics of object types as if they were object types themselves. Consider the following illustration:

> "Elizabeth Taylor's eyes are violet. Violet is a color whose physical wavelength is shorter than that of the other visible colors."

In the first sentence, "violet" was treated as an attribute of the object type "Elizabeth Taylor"; but in the second sentence, "violet" became the subject of its own attributes.

Characterization is a form of abstraction, although a very subtle one. I have added this form to the definition of abstraction. This enhanced definition of abstraction covers most known cases of deriving new object types from old ones.

The products of abstraction have been identified as

- O supertypes
- O subtypes
- O associative types
- O constituent types
- O characteristic types

Characteristic object types are often the most difficult object types to identify and pin down, and nearly always strike the analyst or designer as *too* abstract. Some examples of characteristic object types follow:

- O formal measurements
- O time-dependent information
- O historical information
- O performance analyses
- O projections, correlations, simulations

Budgets, annual reports, and quarterly performance reviews are the kinds of documents that qualify as characteristic object types in complex business information models.

What justifies the introduction of abstract object types into an information model? There are two basic conditions: First, the abstract object type must possess a well-defined function in the system being modeled. (Without a characteristic role or function, the introduction of an abstract object type is gratuitous.) Second, there must exist information that can be attributed to the abstract object type and no other. If there is no information in the domain of information that matches the level of abstraction of the abstract object type, then it is not formally declarable, and its introduction into the model has no justification.

9.1.1 ○ *Final definition: well-defined object type*

All of the decision rules presented in Chapters 5 through 8 operate together to guarantee the meaningfulness of any semantic component of the model. For this reason, we could not have rigorous definitions of our three fundamental concepts before the entire set of rules was developed. Armed with our complete set of rules, we are now ready to proceed to the final definition of the well-defined object type. There are seven conditions that an object type must meet in order to be judged well defined, and which serve as constraints on the modeling process.

DEFINITION (FINAL): **well-defined object type**

a type or class of individual objects is called *well defined* if it meets all seven of the following conditions:

☐   It is possible to state a well-defined function that characterizes the object type uniquely and is preserved as a whole over the model's duration. (Rule 1)

☐   It is possible to identify each unique occurrence of this object type with a unique identifier. (Rule 2)

☐   It is formally declarable in the model. (Rules 3 and 4)

☐   It can be distinguished from other object types in the model. (Rules 5 and 6)

☐   It can be distinguished from other semantic constructs in the model. (Rules 4 and 7)

☐   It can be defined independently of the definition of other object types in the model. (Rule 8)

☐   It is possible to define its attributes and characteristics in the same way for all occurrences over the duration of the model. (Rules 1 and 3)

9.2 ■ Well-defined relationships

A generalized relationship, as we discussed in Chapter 8, is an association between two or more object types, each of which participates in the relationship. Relationships represent interactions between the participating object types. Figure 8.3 shows a relationship (DETER-MINES) between four participating object types (CONTRACT, SALESPER-SON, CONTRACT OFFICER, PROJECT).

A generalized relationship linking an arbitrary number of object types is a natural way to describe interactions between object types in the system being modeled. In the relationship DETERMINES shown in Fig. 8.3, all four object types participate in the contractual process. Generalized relationships permit the analyst to model real-world situations without artificial restrictions.

The generalized relationship is a relatively modern conception in the methods of information modeling. The traditional approach, reflected in most commercially available database management systems, has been to treat all associations in terms of binary relationships.* Although binary relationships are easier to treat than generalized relationships, they tend to fragment and distort models of complex situations in which three or more object types are actively participating. Therefore, generalized relationships are preferable for modeling real-world systems.

Along with the naturalness of generalized relationships comes the complexity of specifying them and using them with precision. As part of the specification process, we ask the following question: "For a relationship between N participating object types, how do we specify the association between the occurrences of some participating object type and occurrences of any other participating object type?"

Determining how to specify this association is very important for setting out a meaningful definition of a relationship. If we cannot say which CONTRACTS are associated with which PROJECTS, CONTRACT OFF-ICERS, and SALESPERSONS, then the relationship DETERMINES is too ambiguous to have meaning in our model.

The method of anchor-point analysis was introduced to solve this problem. In this method, one of the participating object types serves as an anchor, or reference point, for the remaining object types. All occurrences of the remaining object types are referred to the occurrences of the anchor object type. In this way, associations are clearly specified between occurrences of all object types, because all associations can be referred back to a common point.

---

*This method is known as Bachman diagramming, and is used in traditional database design.

9.2.1 ○ *Final definition: well-defined relationship*

Having summarized the logical foundations developed for the concept of the well-defined relationship, we define it here in final form.

DEFINITION (FINAL): **well-defined relationship**

a relationship between N participating object types is called *well defined* if all three of the following conditions are met.

□ It is possible to define the association between the participating object types in terms of the rules, laws, conventions, or policies that govern the interactions between the object types in the system being modeled. (Rule 13)

□ It is possible to state any conditions that determine whether associations hold between specific occurrences of participating object types. (Rule 13)

□ It is possible to associate any occurrence of a participating object type with specific occurrences of any other participating object type unambiguously. (Rule 14)

9.3 ■ Well-formed models

Now that we have defined the conditions that determine whether object types and relationships are well defined, the natural question arises as to whether the entire model is well defined. As we discovered in Chapter 7, there are definite constraints that can be placed on the model to guarantee that it is well formed. The first step is to remove any fuzziness from the information model, by removing fuzzy object types and relationships, and replacing them with well-defined components. Intersections may exist if any data elements in the data dictionary are simultaneously attributable to multiple object types, but are not attributable to any specific object type. Intersections should not be allowed unless these rules are followed.

Another criterion for a well-formed model is that the set of object types should be sufficient to cover the domain of information: No definable and necessary object types should be missing from the model. Conversely, there should be no more object types than are needed to cover the domain of information. If the set of object types abides by the "all that are needed/only those that are needed" rule, then the set is minimally complex.

The attribution rule is formulated to ensure that data elements are attributed to object types only if they truly describe those object types.

Following this rule, data elements should not be misplaced descriptors of the wrong object types, and distribution of data elements throughout the object types of the model should be economical and non-redundant. The rule also implies that there should be no unnecessary overlapping of information in the model. If these conditions are true, then the data structure of the information model is highly modular.

In order to keep the model simple, there should be no nesting: no relationships of relationships or attributes of attributes. Relationships are between object types, not other relationships, and data elements are attributed to object types, not other data elements. In other words, the simplest information model is "flat," with no nested or compounded structures. In technical terms, any system that is free from nested or compounded structures is called a *first-order system*. In order to keep an information model simple, we must require it to be a first-order information model.

Finally, since we have taken care to define each component in terms of the policies, laws, rules, and conventions that govern the system being modeled, it is necessary that all component definitions be mutually consistent and free from contradiction. This guarantees the consistency of the information model.

9.3.1 ○ Final definition: well-formed model

Having summarized its logical foundations, let us proceed to the definition of the well-formed model.

DEFINITION (FINAL): **well-formed model**

an information model, consisting of object types, relationships, operations, regulations, intersections, and attributed data elements, is called *well formed* if the following six conditions are met by the model.

☐   All object types and relationships are well defined.

☐   Intersections exist in the model only if the intersection data cannot be attributed to some well-defined object type by the attribution rule. (Rule 3)

☐   The object set of the model is modular with respect to the distribution of attributes. (Rule 12)

☐   The object set of the model is of minimal complexity. (Rule 11)

☐   The information model is first-order; that is, it contains no nested structures or compounded characteristics. (Rules 7 and 8)

☐   Definitions of model components must be consistent and mutually compatible with the definitions of other components in the model.

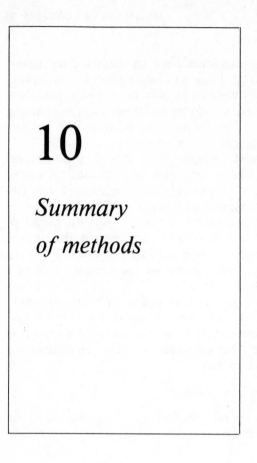

10

Summary of methods

10.1 ■ Philosophy of information modeling

Throughout this monograph, I have been developing fundamental semantic concepts for the analytical portion of information modeling. By analytical, I mean the investigative and model-building components of the total modeling process.

Analysis is a problem-solving process that begins with an unstructured problem domain, and systematically discovers and identifies the central features, components, and issues that govern the problem domain. Analysis is finished when it has *structured* the problem domain. The final structuring of a problem domain is represented by a *completed model* of the domain.

The analytical component is the heart of the entire information modeling process. It presents a method of thinking about how systems are organized; and it presents a procedure for gaining knowledge about the inner workings of the system being studied.

Information modeling has built powerful semantic concepts to assist with the conceptual analysis of the system being studied. The techniques of conceptual analysis enable the analyst to frame questions about the underlying business policy that governs the structure of business systems. Information modeling addresses business policy questions *directly,* rather than indirectly, as in the process of organization-wide data collection or of mathematical dependency analysis.

Information modeling can proceed with a direct analysis of business policy because it uses semantic modeling concepts and an analytical language that permits it to frame questions and express policy at precisely the right level of abstraction.

The remainder of this chapter will focus on the analytical procedures of information modeling, and the structure of the *process* of information modeling as a whole.

10.2 ■ Summary of analytical procedures

Information modeling goes beyond a discussion of technical concepts by offering the system analyst a set of analytical procedures that permit the technical concepts to be put into practice. Five core procedures have been included in this volume:

- O functional analysis
- O scenario analysis
- O transaction analysis
- O abstraction analysis
- O anchor-point analysis

The first three procedures provide general problem-solving strategies for finding and identifying object types, relationships, and operations. Abstraction analysis assists the analyst with the identification of abstract object types. Anchor-point analysis is used to produce a systematic analysis of generalized relationships that are well defined. All five analytical procedures are reviewed below.

10.2.1 O *Functional analysis*

Functional analysis is built on an input-output conceptual model of systems. The black box in the center of the model represents a business function. Using the functional analysis approach, the analyst looks at a business as a large collection of interconnected business processes or functions. The main benefit of this approach is that it can be followed at multiple levels of detail, ranging from very general to very detailed. It permits *functional decomposition*. Model components are identified as follows:

- O   input object types
- O   output object types
- O   input-output relationships
- O   function control entities
- O   operations on parts of the system

10.2.2 O *Scenario analysis*

Scenario analysis views a system as an environment that surrounds some central activity. All participants in the central activity are systematically identified. The procedure uses a deep background approach to system analysis. Model components are identified from

- O   active participants
- O   relationships between active participants
- O   characteristics of participants
- O   organizational structures

10.2.3 O *Transaction analysis*

Transaction analysis is based on an event model of the target system. All changes in the state of the system are studied in terms of the events that bring about state-changes. The state-changes are analyzed into the distinct conditions that affect object types, relationships, and data content. Model components are identified from

- O   transactions affecting the system
- O   operations that compose the transactions
- O   altered object types
- O   altered relationships
- O   altered data elements

10.2.4 O *Abstraction analysis*

Abstraction analysis assists the analyst in defining abstract object types. It applies various decision rules that have been developed as aids to design decision-making. It assists the analyst in identifying

- O   declarable object types
- O   object subtypes
- O   object supertypes
- O   associative object types
- O   characteristic object types

10.2.5 ○ Anchor-point analysis

Anchor-point analysis facilitates analysis of generalized n-ary relationships. By using this procedure, the analyst can arrive at a precise definition of an n-ary relationship that is consistent with known policy, *and* that is well defined. Anchor-point analysis is a new form of combinatorial relational analysis that essentially replaces the intuitive one-to-one, one-to-many approach. It treats

○ generic relationships
○ characteristic relationships
○ selection of anchor object
○ policy definition of relationship
○ anchored binary relationships between participating object types

These five core procedures are components of the analysis phase of information modeling. However, they do *not* exhaust all procedural elements of information modeling. Figure 10.1 illustrates this point by depicting the major phases of information modeling.

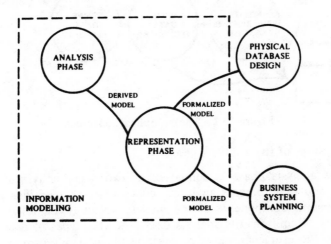

Figure 10.1. The major phases of information modeling.

In the next two sections, the activities of information modeling are further described. First, we will look at the procedural elements of the analysis phase and then at those of the representational phase. Although partially indicated in Fig. 10.1, details of the other phases of information modeling are inappropriate for the level of this book and will not be covered.

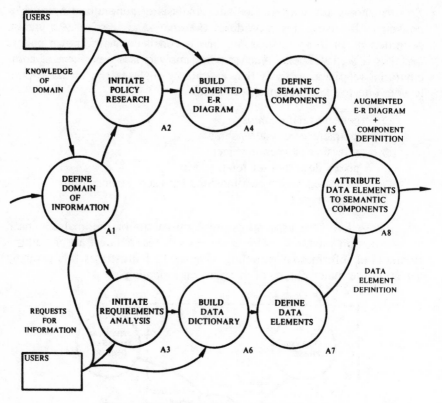

Figure 10.2. Analysis phase, part one.

10.3 ▪ Steps of the analysis phase

The analysis phase is composed of two major activities: policy research and requirements analysis. Policy research identifies the primary semantic components of the model: object types, relationships, operations. Requirements analysis discovers what types of data are required by the user community to describe the semantic components of the model (see Fig. 10.2). Individual steps of the analysis phase are described below with steps A1 through A8 depicted in Fig. 10.2. Figure 10.3 corresponds to steps A9 through A11.

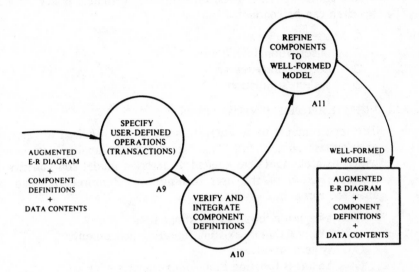

Figure 10.3. Analysis phase, part two.

ANALYSIS PHASE STEPS

A1:   DEFINE DOMAIN OF INFORMATION.

This preliminary step defines the boundaries of the domain of information. This step does *not* derive a detailed specification of the domain of information; that is provided in the completed information model.

A2:   INITIATE POLICY RESEARCH.

Policy research is the fundamental activity of the system analyst in trying to identify the salient entities, relationships, and operations that make up the system being modeled. Definitions of each model component are based on definable, verifiable policy. Policy research can be conducted in

- O   user interviews
- O   observational field work
- O   documentary research
- O   system simulation

A3:   INITIATE REQUIREMENTS ANALYSIS.

The requirements to be analyzed are the data required to describe each system component. This means that the analyst determines which data elements are needed to describe model components to the satisfaction of the user community. Requirements can be analyzed through

- O   user interviews and questionnaires
- O   examination of existing reporting instruments
- O   system simulation
- O   business function input-output analysis

A4:   BUILD AUGMENTED E-R DIAGRAM.

An augmented E-R diagram contains symbols for object types, relationships, and operations. This step is the most rigorously analytical task in information modeling. Policy research is conducted by means of the five core analytical procedures summarized in Section 10.2 (functional analysis through anchor-point analysis). As a result of this step, the majority of the semantic components of the information model have been identified.

A5:   DEFINE SEMANTIC COMPONENTS.

Policy definitions are formulated for object types and relationships in this step. The definitions are written according to rules, as specified in Chapters 4 through 8. The derivation of definitions for user-defined operations is postponed until data elements have been attributed to the object types and relationships in the model.

A6:   BUILD DATA DICTIONARY.

Data elements are collected from user requirements. The set of identified data elements is examined for redundancy and overlap of information content. The final collection of data elements will contain no overlapping information.

A7:   DEFINE DATA ELEMENTS.

The exact information content of each data element is defined in this step. Definitions are policy-dependent.

A8:   ATTRIBUTE DATA ELEMENTS TO SEMANTIC COMPONENTS.

In this step, we apply the attribution rule to object types and relationships. Intersections are considered to be a special case of relationships, and are introduced only if needed. The data content for each component is specified.

A9:   SPECIFY USER-DEFINED OPERATIONS.

Now that object types and relationships possess attributed data elements, it is possible to provide detailed definitions of user-defined operations. The definition for a user-defined operation, such as PLACE ORDER, will make use of the standard operations ADD/DELETE/MODIFY, and the names of participating object types, relationships, and data elements. Allowable operations are specified for each component.

A10:   VERIFY AND INTEGRATE COMPONENT DEFINITIONS.

Since definitions exist for object types, relationships, operations, and data elements, the model can now be verified with the user community. This is an iterative process that continuously refines the policy specifications for each component. Also, disagreements between users as to component definitions can be resolved at this point, and an integrated set of definitions can be derived.

A11:   REFINE WELL-FORMED MODEL.

In this step, the model designer verifies that the object types and relationships are well defined and that the model as a whole is well formed. Changes then are made to assure that the model is well formed.

10.4 ■ Steps of the representation phase

The formal representation of an information model is concerned with finding data structures that represent each component, and with the logical integration of the resulting model. These steps are depicted in Fig. 10.4 and described on the following pages.

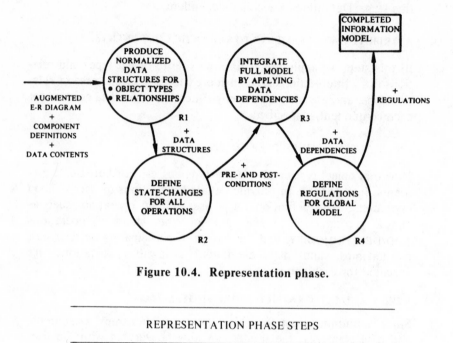

Figure 10.4. Representation phase.

REPRESENTATION PHASE STEPS

R1: PRODUCE NORMALIZED DATA STRUCTURES FOR COMPONENTS.

Data structures for representing components appear simply as hierarchical families of value tables. Each column in a value table is named by a data element name, and each row of the value table represents an occurrence of the semantic component that each value table represents. A well-defined object type is represented by one or more value tables in a hierarchical form. The highest value table contains all single-valued attributes. Lower level value tables contain multi-valued attributes of the object type. Normalization is a design check on the composition of value tables that guarantees that all attributes are placed on the proper access path. Third normal form means that all attributes in a given value table can be accessed solely by the key that accesses that value table and no other. It is a measure of the structural stability of the model representation.

R2: DEFINE STATE-CHANGES FOR ALL OPERATIONS.

Changes of state are specified by pre- and post-conditions attached to operations. Operations are either standard or user-defined. All operations that affect the model (or database) in a significant way are analyzed in terms of the conditions that must exist before the operation takes place (pre-conditions) and the conditions that must hold after the operation takes place (post-conditions). In this way, operational state-changes are defined. Pre- and post-conditions are not necessary for user-defined operations if the user-defined relationship is decomposed into standard operations, and pre- and post-conditions are specified for them.

R3: INTEGRATE MODEL BY DATA DEPENDENCIES.

Data dependencies are logical connections between conditions in the model. The logical connections are of the form "IF condition-1 THEN condition-2." The conditions are defined as data elements having certain values or object occurrences participating in specific relationships. Since model components are defined and data structures assigned to each component, it is the proper time at this step to formulate all relevant data dependencies.

R4: DEFINE REGULATIONS FOR GLOBAL MODEL.

This final step declares the high-level rules and constraints that apply to the information model as a whole.

10.5 ■ Subsequent design issues

An information model is a high-level conceptual model of some real-world system. It is an excellent candidate for use as a logical database design. However, most commercially available database systems place restrictions on the types of logical data structures and relationships that they can implement. These restrictions vary from very little to very great; relational database systems can implement the data structures of object types and relationships directly, with no restrictions, while most other types of database systems *do* have restrictions.

Most of the restrictions involve implementations of relationships. In general, commercially available database systems do not support the concept of a generalized n-ary relationship. Binary relationships are standard in most systems. This does not mean that n-ary relationships are impossible to implement, or that they are not cost-effective to implement. For example, an implementation technique called "inverted list" is widely applied and can be readily used to implement n-ary relationships.

Most probably, the predominance of binary relationships in commercially available systems results from the lack of attention paid to the issue of *logical database design*. While early methods of physical database structuring quickly developed sophistication, logical database design received less attention; virtually no effort was directed toward the rigorous analysis of n-ary relationships. The simplest case, that of binary relationships, was the only relational structure given consideration. We have always had the ability to implement n-ary relationships cost-effectively, but the possibility has never been seriously entertained in logical database design.

Anchor-point analysis has changed the state of the art of logical database design by providing a straightforward procedure for properly defining n-ary relationships with a high degree of reliability and precision. However, the fact remains that most database systems cannot implement n-ary relationships. This means that all n-ary relationships in an information model must be converted to equivalent sets of binary relationships. Of course, the anchor-point analysis approach offers a direct way to do this.

Indeed, a number of steps may be required in order to prepare the database partition of an information model for physical implementation. This set of steps is usually called the *model transformation phase*. Typical steps in the model transformation phase are

(1) conversion of n-ary relationships to sets of binary relationships

(2) mapping of data structures onto the record formatting capabilities of the database system

(3) removal of unique identifiers from those segments whose occurrences can be identified by referring to existing object types

(4) planned introduction of data redundancy to improve processing performance of the physical database

(5) conversion of the networking logical structure of the model to a form compatible with hierarchically organized databases

10.6 ▪ Concluding comments

A subset of model transformations is a set of actions called *design compromises*. At every point of the information modeling procedure, I have emphasized the need for *quality* in the finished model. In fact, the semantic concept of the well-formed model is the formal set of quality constraints that enforces quality control of a completed information model. In actual implementations of the database partition of the model, there may be compelling practical reasons for breaking the quality rules of information modeling. Excessive expense or the need for high performance of the operational database may mean that many of the rules will have to be broken in limited ways. These rule-breaking decisions are the design compromises referred to above. Design compromises should always be made in a careful, premeditated way.

Final Perspective

Information modeling has transformed entity-relationship analysis from a good idea to a complete analytical discipline. Most of the fundamental problems of business system analysis have been addressed and resolved.

Information modeling is a new form of system analysis that complements other forms of system analysis. It is a non-quantitative and descriptive approach to system modeling. It has been developed as a tool for modeling the *business subject matter* of any given enterprise. It does not describe the operational structure or input-output aspects of real-world business systems. It is dynamic, but not procedural in its descriptive content. It is an ideal tool for logical database design, or business problem-solving and business system planning.

The modern system analyst now has the opportunity to fill his or her tool kit with sophisticated tools and methods. The challenge of the coming years will be to apply these new tools to the complex task of organizing the internal operations of the modern enterprise.

Glossary

abstraction

the derivation of a new object type by the processes of observer selection of object characteristics that are preserved as a whole over time, or object normalization of compounded characteristics

abstraction analysis

an analytical procedure that refines and develops the object set of an information model through the application of various modes of abstraction

analysis

a systematic activity that interprets and provides structure for an unstructured problem domain; one of the two major phases of information modeling, it is concerned with policy research and data requirements analysis leading to definitions of the major components of information models

analytical procedure

a multi-step procedure that performs some specific task in the analysis of a real-world system

analytic partition

a portion of the documentation for information models that contains all verbal descriptions of the model components

anchored binary relationship

a relationship between two object types such that the association between the occurrences of the object types is determined by a common association to occurrences of some anchor object; each occurrence of either object type must be associated with some occurrence of the other object type

anchor object an object type that participates in an N-way relationship, and is designated to serve as a reference point for the specification of all combinations of occurrences of all participating object types

anchor-point analysis an analytical procedure that specifies for any n-ary relationship that there are $N(N - 1)/2$ anchored binary relationships associated with the relationship, and consequently that the relationship is well defined

association a form of abstraction that treats a relationship between N object types as an object type in its own right

associative object type an object type that defines a specific relationship between two or more distinct object types; also called an *aggregate*

attribute a named characteristic of the object type such that for any single occurrence of the object type the attribute associates one or more data values with that occurrence

augmented entity-relationship diagram a diagrammatic representation of a system that shows object types (entities), relationships, and operations. The symbols are rectangles for entities, diamonds for relationships, and ellipses for operations

Bachman diagram a diagrammatic representation of a system that shows object types and binary relationships; the symbols are rectangles for entities, arrows for binary relationships

base segment a segment that contains all single-valued attributes of an object type or relationship

business activity a unit business process (billing, general ledger update, and so on) that supports some business function

business entity a component of a business system that plays a specific role in that system

business function a coherent set of business activities that accomplishes a specific business objective (marketing, accounting, and so on)

business subject matter
the definition and specification of all activity making up a given line of business; in general, it is not concerned with the details of the implementation of business activities

characteristic object type
an object type X characterizes an object type Y if X describes Y in some way, specific occurrences of X are attributed to specific occurrences of Y, *and if the non-existence of Y implies the non-existence of X;* an object type is *independent* if it is not a characteristic object type

classification
a form of abstraction that treats a class of object types as an object type in its own right

component
a part of the system being modeled; for information models, the components are object types, relationships, operations, data elements, and regulations

component occurrence
an instance of an object type, relationship, operation, or data element

component type
the class of all occurrences of some object type, relationship, operation, or data element; or a regulation

constituent object type
any object type that relates to other object types by way of some associative object type

database
a systematic and logically organized collection of facts; the logical structure of the data is derived from an information model; a database will store all occurrences of object types, relationships, and data elements

data content
the set of associated data elements that describe object types and relationships

data dependency
a rule that states that if some condition A is true, then some condition B must be true; data dependencies are the chief tools for logical integration of an information model

data dictionary
a document in which descriptive entries for model components are maintained

data elements
    unit facts that describe object types or relationships; data elements are associated with object types and relationships if they describe those components in a meaningful way

data structure
    a hierarchical collection of value tables, such that each value table (segment) has data element names as column headings, and combinations of data element values in each of the rows; there is one base segment (table) and multiple dependent segments

dependent segment
    a segment that contains coordinated multi-valued attributes of an object type or relationship

design compromise
    a structural change in an information model that deliberately breaks the rules of information modeling in order to make a physical implementation more economical to operate

domain of information
    a real-world system, or a bounded portion of that system, that is the subject of an information model; it is represented in an information model by model components, data structures (objects, relationships, and so on), and data elements

entity-relationship analysis
    a form of system analysis that views a system as an abstract collection of entities and relationships; it does not pay attention to processes or inputs and outputs

event model
    a model of some real-world system that views the system in terms of the succession of events that comprise the history of the model

formal declarability
    a set of conditions that justifies the declaration of an object type in the analytic partition of an information model; it must have a well-defined function, a unique identifier, and must contain at least one non-identifier attribute

Formal English	a highly disciplined procedure for verbally describing model components; each statement is composed of three parts: a *prefix* that identifies and quantifies the components referenced in the policy declaration, a *declaration* of known policy governing the referenced components, and a *set of conditions* that qualifies the central policy declaration
functional analysis	an analytical process for the derivation of object types, relationships, and operations based on viewing the domain of information in terms of the input-output characteristics of business functions associated with the domain
functional dependence	a correspondence between two data elements, A and B, such that for any specific value of A there corresponds one and only one value of B; the correspondence is usually defined in terms of a mathematical relation where A and B are represented by columns
functional differentiation	a form of abstraction that treats a subset of occurrences of some object type as an object type in its own right
functional identity	a set of object characteristics, inherent or assigned, that is preserved as a whole over time
fuzzy object	an object for which it is *not* possible to define its role or function in the system being modeled, identify unique occurrences of the object, distinguish it from other types of objects in the model, distinguish it from other semantic constructs in the model (for example, relationships), or define its characteristics in the same way for all occurrences
fuzzy relationship	a relationship for which it is not possible to specify the rules, conventions, or laws that determine the association between participating object types, or specify *unambiguously* how a given occurrence of a participating object type is associated with specific occurrences of other participating object types

generalization an alias for *classification*

graphic partition a portion of the documentation for information models that contains diagrammatic representations of the system being modeled

identification a process of discovering the identities of semantic components in an information model

identifier attribute an attribute X of an object type Y is an *identifier of Y* if there is a unique value of X for each distinct occurrence of Y, *and* there is a public policy that recognizes X as a key of Y; any attribute that is not an identifier attribute is called a *non-identifier attribute*

independent object type an object type that does not characterize any other object type or that can be defined independently of the context of any other object type

information model a representation of some real-world system that identifies the object types, relationships, and operations of that real-world system; it represents object types and relationships by data structures and describes the logical rules that govern the integrity of the model; it is comprised of graphic descriptions, verbal analytic descriptions, and data representations

intersection a data structure composed of a concatenated primary key and intersection data; the primary key is a concatenation of the primary keys of the object types on which the attributes depend; each attribute is value-dependent on the full combination of the primary keys

intersection data one or more data elements taken from the data dictionary that are partially attributable to two or more object types in the model; the values of these data elements are value-dependent on the *combination of* primary keys of two or more object types in the model

inverted list a technique for indexing data structures in which the values of data elements are used as pointers to access or associate their respective data structures

mathematical dependency analysis   a set of procedures for determining precise, quantifiable correspondences between logically related pairs of data elements

mathematical relation   a two-dimensional table of values, with any number of columns or rows. For any column in the table, the values in the column are selected from a pre-specified set of values; each row in the table represents some occurrence of a model component

minimal complexity   a condition that guarantees that an information model contains *all* the object types needed to describe a domain of information, and *only* the object types needed to describe the domain

model   a representation of a real-world system; in general, there is a one-to-one correspondence between the components of a model and some subset of the elements of the real-world system

model transformation   an alteration of the structure of an information model to permit it to be implemented in a physical database system

modularity   a condition that guarantees that all data elements are attributed to the components they describe, and *only* the components they describe

non-identifier attribute   see *identifier attribute*

normalization   a process for converting data structures to third normal form; not to be confused with database design procedures

object characteristic   a distinguishing feature that permits an observer to identify an object as belonging to a particular class or type; the six kinds of object characteristics are *purpose* (the reason, or set of reasons, for the presence of the object in the system being modeled), *defining properties* (the essential observable features of the object associated with all occurrences of the object), *object effect on the system* (the observable effect of the action of the object on the system being modeled), *system effect on the object* (the observable effect of the action of the system on the object), *association with other objects* (the associations abstracted from the object's interaction with other objects or participation in groups), and *behavior pattern* (the observable pattern of the object's behavior within the system being modeled)

object function
a set of characteristics of some object, inherent or assigned, that is of interest to an observer of the system being modeled; the set may contain all or part of the totality of object characteristics

object normalization
a process of converting compounded characteristics into object types in an information model; it is used to eliminate nested structures in an information model

object subtype
a distinguished category, or subset, of all individuals represented by the object type; it is generally derived by functional differentiation of the original object type

object supertype
any object type that includes it as a subtype; generally derived by a classification of the original object type

object type
a class of individuals that is characterized by a distinct functional identity, which is preserved as a whole over time in the system being modeled

operation
an action that changes the state of the system; can be identified with business transactions and events

partition
a portion of documentation for an information model, or the data representation of an information model

policy
the body of rules, laws, and conventions that govern the structure, behavior, and event history of a real-world system; policy may be explicitly written, or encapsulated in the practices that are observed in the daily operation of the system

policy research   an analytical procedure for discovering the business subject matter of some domain of information; the business subject matter is defined in terms of known business policy

post-condition   a condition that *must* obtain in an information model after an operation has been performed

pre-condition   a condition that *must* obtain in an information model in order for an operation to be performed

primary key   a data element that contains a unique value for each distinct occurrence of the object type; a preferred unique identifier

problem domain   a real-world system, or portion of that system, that is the analytical process

regulation   a rule that governs the content, structure, integrity, and operational activity of the model; applies to the model as a whole

relational table   consists of unique identifiers for a set of associated object types; there is one column for each object type participating in the relationship; an object type may occur in more than one column of the table; all entries in a given column are drawn from the domain of unique identifiers of the object type that is represented by that column; each row defines an occurrence of the relationship, which associates distinct occurrences of each of the participating object types; a relational table is complete if all entries in the table have non-null values

relationship   an association defined between occurrences of two or more object types

representation   a major phase in the process of information modeling that is primarily concerned with the formulation of data representations for components in the model

requirements analysis   a set of procedures to define the required data content of a given domain of information

scenario analysis   an analytical process used for the initial identification of object types and relationships in an information model; it uses a deep-background approach to the discovery process

segment   a two-dimensional table of values in which the columns represent attributes (data elements) that comprise the segment, and the rows represent occurrences of the model component associated with the segment

semantic concept   a term used in information modeling to denote a component of the system being modeled; examples are object types, relationships, operations; these terms apply to components of real-world systems whose meaning is assigned to these terms by convention

semantics   the preferred, or conventional, use of a term or known concept; a body of methods and conventions for assigning meaning to the terms of system analysis

standard operation   a primitive operation on model components that is *not* user-defined

state   a set of conditions obtaining in a real-world system at any point in time

state-transition analysis   a form of system analysis that views a system as a collection, or family, of interchangeable states; perhaps the most abstract form of system analysis

system   a totality that emerges from the net effect of interaction between its elements

system analysis   a set of procedures for determining the content, structure, and behavior of the elements that make up a system, and the system considered as a whole

system modeling   a set of procedures for producing a standardized specification of a system being analyzed; the model is a representation of the original system

third normal form   a logical constraint on the design of data structures; it states that all attributes contained in a given segment are functionally dependent on the unique identifier of the segment, the whole identifier, and nothing but the identifier

transaction analysis   an analytical process that identifies all operations that affect an information model by cataloging all business events and transactions that pertain to a given domain of information

unique identifier   a data element that possesses a unique value for each distinct occurrence of the segment to which it belongs

user-defined operation   a state-change action that is defined by the user and is decomposable into a combination of standard operations

value dependence   if A and B are two attributes of object type X, and A is single-valued for each distinct occurrence of X, then B is value-dependent on A if for each distinct value of A there is a finite, unique set of B values that corresponds to it

well-defined function   a function for which its purpose, defining properties, and some combination of the remaining object characteristics can be defined and will remain valid throughout the lifetime of the object types within the system being modeled

well-defined object type   a type or class of individual objects that meets all of the following conditions: it has a well-defined function, which can be stated, that characterizes the object type uniquely and is preserved as a whole over the model's duration; each unique occurrence can be identified with a unique identifier; it is formally declarable in the model; it is distinguishable from other object types and other semantic constructs in the model; it can be defined independently of other object types in the model; and its attributes and characteristics are definable in the same way for all occurrences over the duration of the model

well-defined relationship

a relationship between N participating object types in which it is possible to define the association between the participating object types in terms of the rules, laws, conventions, or policies that govern the interactions between object types in the system being modeled; it is possible to state any conditions that determine whether associations hold between specific occurrences of participating object types; and it is possible to associate any occurrence of a participating object type with specific occurrences of any other participating object types unambiguously

well-formed model

an information model consisting of object types, relationships, operations, regulations, intersections, and attributed data elements in which the following conditions are met: all object types and relationships are well defined; intersections exist only if the intersection data cannot be attributed to some well-defined object type; the object set is modular and minimally complex; the model is first-order; and model component definitions are consistent and mutually compatible with the definitions of other components in the model

Bibliography

Birkhoff, G., and T.C. Bartee. *Modern Applied Algebra.* New York: McGraw-Hill, 1970.

Chen, P.P. "The Entity-Relationship Model—Toward a Unified View of Data." *ACM Transactions on Database Systems,* Vol. 1, No. 1 (March 1976), pp. 9-36.

————, ed. *Proceedings of the International Conference on Entity-Relationship Approach to Systems Analysis and Design,* Los Angeles, 1979.

Codd, E.F. "Extending the Database Relational Model to Capture More Meaning." *ACM Transactions on Database Systems,* Vol. 4, No. 4 (December 1979), pp. 397-434.

Copi, I.M. *Symbolic Logic,* 5th ed. New York: Macmillan, 1979.

Date, C.J. *An Introduction to Database Systems,* 2nd. ed. Reading, Mass.: Addison-Wesley, 1977.

Davis, G.B. *Management Information Systems.* New York: McGraw-Hill, 1974.

Delobel, C. "Normalization and Hierarchical Dependencies in the Relational Data Model." *ACM Transactions on Database Systems,* Vol. 3, No. 3 (September 1978), pp. 201-22.

Einstein, A. *The Meaning of Relativity,* 5th ed. Princeton: University Press, 1974.

Flavin, M. "A Psychological View of Object Conceptualization." Unpublished paper, January 1981.

Kent, W. "Limitations of Record-Based Information Models." *ACM Transactions on Database Systems,* Vol. 4, No. 1 (March 1979), pp. 107-31.

Schevermann, P., G. Shiffner, and A. Weber. "Abstraction Capabilities and Invariant-Properties Modeling Within the Entity-Relationship Approach." *Proceedings of the International Conference on Entity-Relationship Approach to Systems Analysis and Design,* Los Angeles.

Schmid, H.A., and J.R. Swenson. "On the Semantics of the Relational Data Model." *Proceedings of the 1975 ACM SIGMOD Conference on the Management of Data.* New York: Association for Computing Machinery, 1975, pp. 211-23.

Smith, J.M., and D.C.P. Smith. "Database Abstractions: Aggregation." *Communications of the ACM,* Vol. 20, No. 6 (June 1977), pp. 405-13.

_____. "Database Abstractions: Aggregation and Generalization," *ACM Transactions on Database Systems,* Vol. 2, No. 2 (June 1977), pp. 105-33.

Von Bertalanffy, L. *General System Theory.* New York: George Braziller, 1968.

Zaniolo, C., and M.A. Melkanoff. "On the Design of Relational Database Schemata." *ACM Transactions on Database Systems,* Vol. 6, No. 1 (March 1981), pp. 1-47.

Index